Introduction to Machine Learning in the Cloud with Python

Pramod Gupta • Naresh K. Sehgal

Introduction to Machine Learning in the Cloud with Python

Concepts and Practices

 Springer

Pramod Gupta
NovaSignal
San Jose, CA, USA

Naresh K. Sehgal
NovaSignal
Santa Clara, CA, USA

ISBN 978-3-030-71272-3   ISBN 978-3-030-71270-9   (eBook)
https://doi.org/10.1007/978-3-030-71270-9

This Springer imprint is published by the registered company Springer Nature Switzerland AG
The registered company address is: Gewerbestrasse 11, 6330 Cham, Switzerland

We dedicate this book to Prof. Pramod Chandra P. Bhatt, without his relentless guidance and inspiration, this would have remained only an imagination.

Foreword

The invention of the cloud brought massive advancement in computer architecture. Seemingly overnight the delivery and consumption of services were transformed. Cloud computing has been an accelerant to the pace of technology adoption. In fact, it is cloud computing that has enabled artificial intelligence to move from the world of sci-fi to mainstream consumption. The cloud delivers the massive data storage and compute capacity needed for cost-effective analytics and insights.

As the promises of reduced time to deployment, lower cost of operation, and increased accessibility have been realized, adoption of the cloud has moved from novel to normal. And yet the theoretical understanding of the underlying technology remains locked in the brains of a relatively small group. The availability of cloud and AI curriculum in all educational settings needs to match the availability of cloud and AI services if the current pace of innovation is to continue.

For those looking to gain insight into the complex algorithms of artificial intelligence and the cloud architecture that fuels them, this book fills the void with highly valuable instruction.

NovaSignal, Los Angeles, CA, USA   Diane Bryant
January 18, 2021

Preface

The idea for this book came from our lead coauthor, Pramod Gupta, who has been teaching Data Sciences and related classes at the University of California, Santa Cruz Extension, and most recently at the University of California, Berkeley, for several years. Prior to that his hands-on experience in the industry uniquely qualifies him to write the AI and ML parts of this book. Pramod had met with Naresh K. Sehgal decades ago during their undergraduate studies at Punjab Engineering College, Chandigarh, in India. Since then, Naresh's career path took him through Chip design and later on a journey through Cloud Computing. The amalgamation of their respective work experiences has resulted in this book in your hands. It would not have been possible without the guidance and inspiration of Prof. PCP Bhatt, to whom the authors would like to dedicate this book.

It starts with an introduction to Machine Learning in Chap. 1 and laying down a deeper foundation of ML algorithms in Chap. 2. Then Chap. 3 serves as a bridge to the Cloud, making a case for using the abundant compute and storage availability for training and inference purposes in the context of Deep Learning. Chap. 4 further explains some basic concepts of Cloud Computing with emphasis on the key characteristics that differentiate it from enterprise computing, or even running AI algorithms on a laptop. Chapter 5 expands the usage of Cloud for Machine Learning by enumerating its data pipeline stages. Chapter 6 touches on a very important aspect of security in the Cloud for AI and ML algorithms as well as datasets. Chapter 7 delves into some practical aspects of running ML in Amazon's Cloud setup. Chapter 8 gives an example of using Cloud for health care-based AI and ML solutions. Lastly, in Chap. 9, we look at efforts underway to speed up AI and ML using various hardware-based solutions. No engineering book can be complete without some practical problems and solutions. To meet that expectation, we present three real-life projects that Pramod's students had implemented using Python in Appendices A through C. For each of these, migration of these projects' code to a commercial Public Cloud is illustrated for the reader to practice. Appendix D has solutions to various Points to Ponder, which were posed at the end of each chapter. The motivation here is for the reader to think and then compare one's own answers with our proposed solutions. It can also be the basis of discussion in a classroom setting. The book wraps up with additional questions in Appendix E, the answers to which we leave for the readers to complete.

As with any major project, writing this book took months of planning and over a year to execute it. Even though only two coauthors are listed, there was a major contribution by Prof. PCP Bhatt, who met and reviewed our progress every week for over a year. In addition, we are thankful to our colleague, Aditya Srinivasan, who wrote two sections in Chap. 8 on Multi-Cloud Solutions and UCSD Antibiogram case study. We also used coding examples of several students from Pramod's classes in the Appendices. Naresh picked some ideas and sections from his earlier books with Dr. John M. Acken and Prof. Bhatt on Cloud Computing and Security. NovaGuide View application developed by Shiv Shankar has been used to illustrate NovaSignal's growing presence in the Cloud. We are very grateful to Diane Bryant, the CEO of NovaSignal, for giving us an opportunity to complete this book under her leadership and writing the foreword. Needless to say, several other resources and sites were used to learn and leverage educational material that has been duly acknowledged in our reference sections.

We sincerely hope that readers will have as much fun reading it as we had writing the varied material in this book. We accept ownership for all the mistakes in this book, but please do not miss sending these to us for corrections, in addition to any suggestions for a future edition. If you include at least part of the sentence the error appears in, that makes it easy for us to search. Page and section numbers are fine, too. Thanks!

San Jose, CA, USA   Pramod Gupta
Santa Clara, CA, USA   Naresh K. Sehgal

About the Book

Objective: The purpose of this book is to introduce machine learning and cloud computing, both from a conceptual level and its usages with underlying infrastructure. The focus areas of this book include best practices for using AI and ML in a dynamic infrastructure with cloud computing and high security.

Target audiences are:

1. Senior UG students who have studied programming languages and operating systems.
2. Senior UG and PG students in software engineering or information technology disciplines.
3. SW developers engaged in migrating in-house ML applications to public cloud.
4. Information technology managers for improving AI/ML performance in the cloud.
5. Professionals who want to learn ML and cloud and the technologies behind them.

Level of the book: The book is aimed at senior UG or first semester of PG in software engineering or IT systems.

Contents

Acronyms

AI	Artificial Intelligence
CPS	Cyber Physical Systems
CPU	Central Processing Unit
DDOS	Distributed Denial of Service
DL	Deep Learning
DOS	Denial of Service
GPU	Graphics Processing Unit
IaaS	Infrastructure as a Service
IoT	Internet of Things
ML	Machine Learning
NIST	National Institute of Standards and Technology
PaaS	Platform as a Service
SaaS	Software as a Service
TPU	Tensor Processing Unit
VLIW	Very Long Instruction Word
VPU	Vision Processing Unit
WSE	Wafer Scale Integration

Part I
Concepts

Chapter 1
Machine Learning Concepts

Over the last decade, machine learning (ML) has been at the core of our journey toward achieving larger goals in artificial intelligence (AI). It is one of the most influential and important technologies of the present time. ML is considered an application of AI, based on an idea that given sufficient data, machines can learn the necessary operational rules. It impacts every sphere of human life as new AI-based solutions are being developed.

Recently, machine learning has given us practical speech recognition, effective web search, and a vastly improved understanding of the human genome. Machine learning is so pervasive today that one probably uses it dozens of times daily without realizing it. There is no doubt, ML will continue to make headlines in the foreseeable future. It has the potential to improve as more data, powerful hardware, and newer algorithms continue to emerge. As we progress in the book, we will know that ML has a lot of benefits to offer.

The rate of development and complexity of the field make it difficult even for the experts to keep up with new technique. It can therefore be overwhelming for the beginners. This provided sufficient motivation for us to write this text to offer a conceptual-level understanding of machine learning and current state of affairs.

1.1 Terminology

- *Dataset*: The starting point in ML is a dataset, which contains the measured or collected data values represented as numbers or text, a set of examples that contain important features describing the behavior of the problem to be solved. There is one important nuance though: if the given data is noisy, or has a low signal to noise ratio, then even the best algorithm will not help. Sometimes it is referred to as "garbage in – garbage out." Thus, we should try to build the dataset as accurately as possible.

© The Author(s), under exclusive license to Springer Nature Switzerland AG 2021
P. Gupta, N. K. Sehgal, *Introduction to Machine Learning in the Cloud with Python*,
https://doi.org/10.1007/978-3-030-71270-9_1

- *Features/attributes*: These are also referred to as parameters or variables. Some examples include car mileage, user's gender, and a word's frequency in text, in other words, properties/information contained in the dataset that helps to better understand the problem. These parameters or features are the factors for a machine to consider. These parameters are used as input variables in machine learning algorithms to learn and infer, to be able to take an intelligent action. When the data is stored in tables, it is simple to understand, with features as column names. Selecting the right set of features is very important which will be considered in a later chapter. It is the most important part of a machine learning project's process and usually takes much longer than all other ML steps.
- *Training data*: ML model is built using the training data. This is the data that has been validated and includes desired output. The output or results are generally referred as the labels. The labeled training data helps an ML model to identify key trends and patterns essential to predicting the output later on.
- *Testing data*: After the model is trained, it must be tested to evaluate how accurate it is. This is done by the testing data, where the ML-generated output is compared to the desired output. If both match, then the tests have passed. It is important for both the training and testing datasets to resemble the situations that the ML algorithms will encounter later in the field. Think of a self-driven car's ML model, which has never seen a stop sign during its training phase. Then it will not know how to react when one is seen on actual drive later on.
- *Model*: There are many ways to solve a given problem. The basic idea is building a mathematical representation that captures relationships between the input and output. In other words, it is a mapping function from input to output. This is achieved by a process known as training. For example, logistic regression algorithm may be trained to produce a logistic regression model. The method one chooses will affect the precision, performance, and complexity of the ML model.

To sum up, an ML process begins by inputting lots of data to a computer, then by using this data, the computer/machine is trained to reveal the hidden patterns and offer insights. These insights are then used to build an ML model, by using one or more algorithms to solve other instances of the same problem.

Let us take an example on the following dataset:

Outlook	Temperature	Humidity	Windy	Class
Sunny	Hot	High	False	N
Sunny	Hot	High	True	N
Overcast	Mild	Normal	False	Y
Rain	Cool	Normal	True	Y
Rain	Cool	High	True	N
Overcast	Mild	High	False	Y
Sunny	Hot	Normal	True	Y
Overcast	Mild	High	True	N
Rain	Hot	Normal	False	N

In the above example, there are five features (i.e., Outlook, Temperature, Humidity, Windy, and Class). There are nine observations or rows. In this example, Class is the target or desired output (i.e., to go for a play or no play), which ML algorithm wants to learn and predict for the unseen new datasets. This is a typical classification problem. We will discuss the concept of various tasks performed by ML later in this book.

1.2 What Is Machine Learning?

To demystify machine learning, and to offer a learning opportunity for those who are new to this domain, we will start by exploring the basics of machine learning and the process involved in developing a machine learning model. Machine learning is about building programs with tunable parameters, which are adjusted automatically. The goal is to improve the behavior of an ML model by adapting to previously seen data.

Machine learning is a subfield of artificial intelligence (AI). ML algorithms are the building blocks to make computers learn and act intelligently by generalizing, rather than just storing and retrieving data items like a database system.

While the field of machine learning has not been explored until recently, the term was first coined in 1959 [1]. Most foundational research was done through the 1970s and 1980s. Popularity of machine learning today can be attributed to the availability of vast amounts of data, faster computers, efficient data storage, and evolution of newer algorithms.

At a higher level, machine learning (ML) is the ability of a system to adapt to new data. The learning process advances through iterations offering better quality of response. Applications can learn from previous computations and transactions, by using "pattern recognition" to produce reliable and better informed results.

Arthur Samuel, a pioneer in the field of artificial intelligence, coined the term "Machine Learning" in 1959 while at IBM [1]. He defined machine learning as a "Field of study that gives computers the capability to learn without being explicitly programmed."

In a layman's words, machine learning (ML) can be explained as automating and improving the learning process of computers based on experiences, without explicit programming. The basic process starts with feeding data and training the computers (machines). This is achieved by feeding data to an algorithm to build ML models. The choice of algorithm depends upon the nature of task. The machine learning algorithms can perform various tasks using methods such as classification and regression.

Machine learning algorithms can identify patterns in the given data and build models that capture relationships between input and output. This is useful to predict outcome for a new set of inputs without explicit pre-programed rules or models.

1.2.1 Mitchell's Notion of Machine Learning

Another widely accepted definition of machine learning was proposed by the computer scientist Tom M. Mitchell [2]. His definition states that "a machine is said to learn if it is able to take experience and utilize it such that its performance improves upon similar experiences in the future." His definition says little about how machine learning techniques actually learn to transform data into actionable knowledge.

Machine learning also involves study of algorithms that improve a defined category of tasks while optimizing a performance criterion of past experiences. ML uses data and past experiences to realize a given goal or performance criterion.

Most desirable property of machine learning algorithms is the generalization, i.e., a model should perform well on the new or unseen data. The real aim of learning is to do well on test data that was not known during learning or training. The objective of machine learning is to model the true regularities in a data and to ignore the noise in the data.

1.3 What Does Learning Mean for a Computer?

A computer program is said to learn from experience E with respect to some class of tasks T and performance measure P, if its performance at tasks T, as measured by P, improves with experience E. It can be used as a design tool to help us think about which data to collect (E), what decisions software needs to make (T), and how to evaluate its results (P).

Example: playing tennis
  $E =$ the experience of playing many games of tennis
  $T =$ the task of playing tennis
  $P =$ the probability that the program will win the next game

1.4 Difference Between ML and Traditional Programming

- *Traditional programming*: Feed in data and a program (logic), run it on a machine, and get the output.
- *Machine learning*: Feed in data and its corresponding observed output, run it on machine during learning (training) phase. Then the machine generates its own logic, which can be evaluated during testing phase, as shown in Fig. 1.1.

Fig. 1.1 Basic differences between traditional programming and machine learning

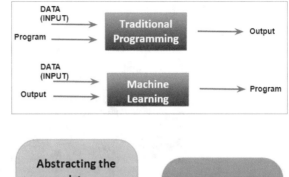

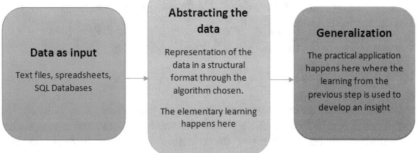

Fig. 1.2 Basic learning process

1.5 How Do Machines Learn?

Regardless of whether the learner is a human or a machine, basic learning process is similar to that shown in Fig. 1.2. It can be divided into three components as follows:

- *Data input*: It comprises observations, memory storage, and recall to provide a factual basis for further reasoning.
- *Abstraction*: It involves interpretation of data into broader representations.
- *Generalization*: It uses abstracted data to form a basis for insight and taking an intelligent action.

1.6 Steps to Apply ML

The machine learning process involves building a predictive model that can be used to find a solution for the given problem. Following steps are used in developing an ML model, as shown in Fig. 1.3.

1. *Problem definition*: This is an important phase as the choice of the machine learning algorithm/model will depend on the problem to be solved. The problem is defined only after the system has been studied well. For example, it may use classification or regression. In particular, the study will be designed to understand

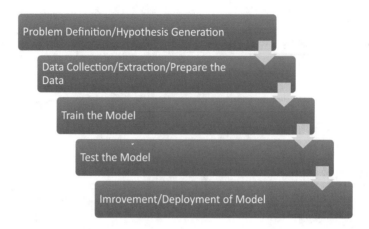

Fig. 1.3 Machine learning process

the principles of its behavior in order to make predictions or to make choices (defined as an informed choice). The definition step and the corresponding documentation (deliverables) of the scientific problem or business are both important to focus the analysis on getting results.

2. *Data collection/data extraction*: The next stage for machine learning model is a dataset. This step is the most important and forms the foundation of the learning. The predictive power of a model depends not only on the quality of the modeling technique but also on the ability to choose a good dataset upon which to build the model. So, search for the data, its extraction, and subsequent preparation related to data analysis because of their importance in the success of the results. Input data must be chosen with the basic purpose to build a predictive model, and its selection is crucial for the success of the analysis as well. Thus, a poor choice of data, or performing analysis on a data set that is not representative of the system, will lead to models that will deviate from the system under study. Better variety, density, and volume of relevant data will result in better learning prospects for the machine learning.

3. *Prepare the data*: Once data has been selected and collected, the next stage is to make sure that the data is in proper format and of good quality. As mentioned earlier, the quality of data is important for predictive power of machine learning algorithms. One needs to spend time determining the quality of data and then take steps for fixing issues such as missing data, inconsistent values, and treatment of outliers. Exploratory analysis is one method to study the nuances of data in details, thereby burgeoning the relevant content of the data. The quality of data is very important for the performance of machine learning algorithms.

4. *Train the algorithm*: By the time the data has been prepared for analysis, one is likely to have a sense of what one hopes to learn from the data. A specific machine learning task will result in the selection of an appropriate algorithm. This algorithm will represent data in the form of a model. This step involves choosing the appropriate algorithm and representation of data in the form of the

model. The cleaned-up data is split into two parts: train and test; the first part (training data) is used for developing the model, and the second part (test data) is used as a reference. The proportion of data split depends on the prerequisites such as the number of input variables and complexity of the model.

5. *Test the algorithm*: Each machine learning model results in a biased solution to the learning problem, so it is important to evaluate how well the algorithm is learned. Depending on the type of model used, one can evaluate the accuracy of the model using a test dataset or may need to develop measures of performance specific to the intended application. To test the performance of the model, the second part of the data (test data) is used. This step determines the precision of the choice of the algorithm based on the desired outcome.

6. *Improving the performance*: A better test to check the performance of a model is to observe its performance on the data that was not used during building the model. If better performance is needed, it becomes necessary to utilize more advanced strategies to augment the performance of the model. This step may involve choosing a different model altogether or introducing more variables to augment the accuracy. Hence, significant amount of time needs to be spent in data collection and preparation. One may need to supplement with additional data or perform additional preparatory work as was described in step 2 of this process.

7. *Deployment*: After the above steps are completed, if the model appears to be performing satisfactorily, it can be deployed for the intended task. The successes and failures of a deployed model might even provide additional data for the next generation of model.

The above steps 1–7 are used iteratively during the development of an algorithm.

1.7 Paradigms of Learning

Computers learn in many different ways from the data depending upon what we are trying to accomplish. There is "No Free Lunch Theorem" famous in machine learning. It states that there is no single algorithm that will work well for all the problems. Each problem has its own characteristics/properties. There are lots of algorithms and approaches to suit each problem with its individual quirks. Broadly speaking, there are three types of learning paradigms:

- Supervised learning
- Unsupervised learning
- Reinforcement learning

Each form of machine learning has differing approaches, but they all follow an underlying iterative process and comparison of actual vs. desired output, as shown in Fig. 1.4.

Fig. 1.4 Three types of learning paradigms

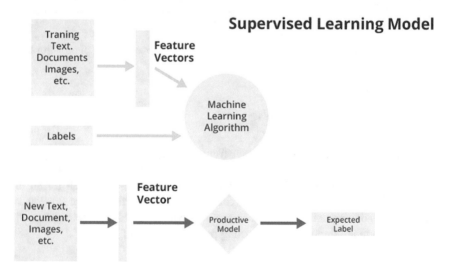

Fig. 1.5 A supervised learning model

1.7.1 Supervised Machine Learning

Supervised learning, as shown in Fig. 1.5, is the most popular paradigm for machine learning. It is very similar to teaching a child with the use of flash cards. If you are learning a task under supervision, someone is judging whether you are getting the right answers. Similarly, supervised learning means having a full set of labeled data while training an algorithm. Fully labeled means that each observation in the dataset is tagged with the answer that the algorithm should learn. Supervised learning is a form of machine learning in which input is mapped to output using labeled data, i.e., input–output pairs. In this case, we know the expected response, and the model is trained with a teacher. In this type of learning, it is imperative to provide both inputs and outputs to the computer for it to learn from the data. The computer generates a function based on the data that can be used for the prediction of unseen data. Once trained, the model will be able to observe a new, never-seen-before example and predict an outcome for it. The trained model no longer expects the target. It will try to predict the most likely outcome from a new set of observations. The solution can use classification or regression depending on the type of the target.

Depending upon the nature of the target, supervised learning can be useful for classification as well as regression type of problems.

If target y has values in affixed set of categorical outcomes (e.g., male/female, true/false), the task to predict y is called classification.

If target y has continuous values (e.g., to represent a price, a temperature), the task to predict y is called regression.

1.7.2   Unsupervised Machine Learning

Unsupervised learning is the opposite of supervised learning. It uses no labels. Instead, the machine is provided with just the inputs to develop a model, as shown in Fig. 1.6. It is a learning method without target/response. The machine learns through observations and finds structures in the data. Here the task of machine is to group unsorted information according to similarities, patterns, and differences without any prior training. Unlike supervised training, no teacher is provided that means no training will be given to the machine. Therefore, the machine is restricted to find hidden patterns in unlabeled data. An example would be to perform customer segmentation or clustering. What makes unsupervised learning an interesting area is that an overwhelming majority of data in our world is unlabeled. Having intelligent algorithms that can take terabytes of unlabeled data and make sense of it are a huge source of potential profit in many industries. This is still an unexplored field of machine learning, and many big technology companies are currently researching it.

1.7.3   Reinforcement Machine Learning

Reinforcement learning allows machine to automatically determine the ideal behavior within a specific context, in order to maximize its performance. Reinforcement learning is looked upon as learning from mistakes as shown in Fig. 1.7. Over time,

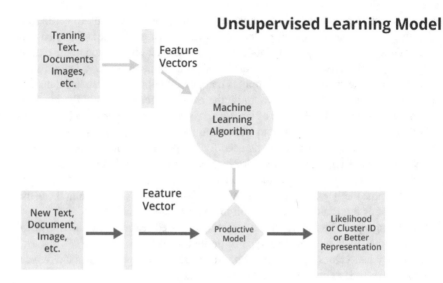

Fig. 1.6 An unsupervised learning model

Fig. 1.7 Reinforcement learning

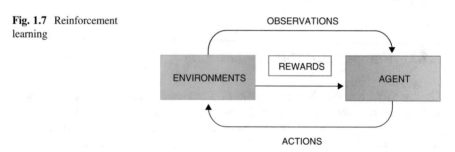

learning algorithm learns to make fewer mistakes than it used to. It is very behavior driven.

This learning paradigm is like a dog trainer, who teaches the dog how to respond to specific signs, like catch a ball, jump, or anything else. Whenever the dog responds correctly, the trainer gives a reward to the dog, which can be a "bone or a biscuit."

Reinforcement learning is said to be the hope of artificial intelligence because the potential it possesses is immense for many complex real-life problems, such as self-driving cars.

1.7.3.1   Types of Problems in Machine Learning

As depicted in Fig. 1.8, there are three main types of problems that can be solved using machine learning:

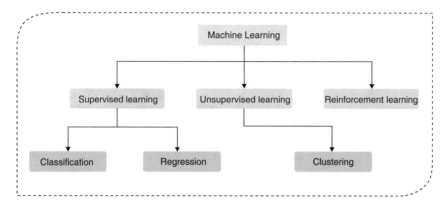

Fig. 1.8 Types of problems solved using machine learning

Fig. 1.9 Classification
using machine learning

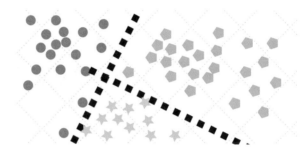

Classification problem: Classification is the process of predicting the class of a given data points. Classification predictive modeling is the task of approximating a mapping function from input variables to discrete output variables, e.g., spam detection in emails and credit card fraud. In these cases, we try to draw a boundary between different classes as shown in Fig. 1.9. A classifier utilizes some training data to understand how given input variables relate to the class. The dataset may simply be bi-class (e.g., is incoming mail a spam or non-spam?) or it may be multi-class (e.g., health of a patient). Some other examples of classification problems are speech recognition, fraud detection, documents classification, etc. There are various ML algorithms for classification that will be discussed later.

Regression problem: Regression is the task of predicting the value of a continuously varying variable (e.g., a sale price of a house or a height of a tree) given some input variables (aka the predictors, features, or regressors). A continuous output variable is a real-value, such as an integer or floating-point value. These are often quantities such as the amounts and sizes. It tries to model data distribution with the best line/hyper-plane which goes through the points as shown in Fig. 1.10, Regression is based on a hypothesis that can be linear, polynomial, nonlinear, etc. The hypothesis is a function that is based on some hidden parameters and the input values.

Fig. 1.10 Regression using machine learning

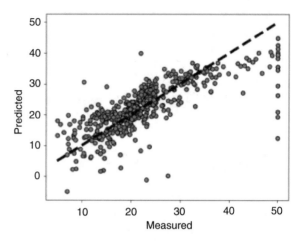

Fig. 1.11 Clustering

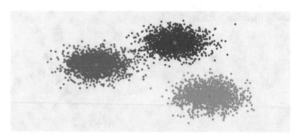

Clustering: This type of problem involves assigning the input into two or more clusters based on similarity as shown in Fig. 1.11, for example, clustering customers into similar groups based on their spending habits, age, geography, items they buy, etc. This is unsupervised learning as there is no target available in advance.

Figure 1.12 sums up the differences between regression, classification, and clustering.

1.8   Machine Learning in Practice

Machine learning algorithms are a small part of practices by a data analyst or data scientist to do machine learning. In reality, the actual process often looks like:

- Start loop

  - *Understand the domain, prior knowledge, and goals.* Start by talking to the domain experts. Often the goals are unclear. One may have to try multiple approaches before starting to implement.
  - *Data integration, selection, cleaning, and pre-processing.* This is often the most time-consuming part. It is important to have high-quality data. The more

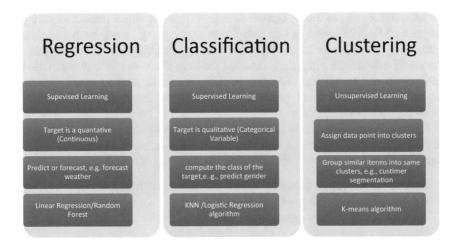

Fig. 1.12   Regression vs. classification vs. clustering

data one has, more work may be required because the data can be noisy, remember GIGO (garbage in, garbage out).

- *Learning models*. This is an exciting phase with availability of many tools to experiment with.
- *Interpreting results*. Sometimes it does not matter how a model works as long as it delivers the results. Some domains require that the model is understandable, so we need to be prepared to be challenged by the experts.
- *Consolidating and deploying discovered knowledge*. The majority of projects that are successful in the lab may not be used in practice. In such cases, compare the model's output with desired results.

• End loop

Clearly it is not a one-shot process, but an iterative cycle. It also explains that learning happens by observing. The reason is to learn from the gaps between actual and desired results. The loop iterates until we get a model and the results that can be used in practice. Also, incoming data may change, requiring a new loop.

1.9   Why Use Machine Learning?

It is important to remember that machine learning (ML) does not offer solutions to every type of problem at hand. There are many cases for which solutions can be developed without using ML techniques. The defining characteristic of a rule-based machine learning algorithm is the identification and utilization of a set of relational rules, which collectively represents the knowledge captured by the system. For example, we do not need ML if we can determine a target value by using pre-established simple rules, computations, or predetermined steps that can be programmed without needing any data-driven learning, e.g., a rule such as {onion, lettuce} → {Bread} found in the sales data of a supermarket indicates that if customers buy onions and lettuce together, they are likely to also buy bread. Such information can be used as the basis for decisions about marketing and planning activities, e.g., promotional products, co-location of items stocked on nearby shelves.

Machine learning is usually used for complex tasks or problems involving a large amount of data and with lots of incoming features. For example, machine learning is a good option if we need to manage the following types of situations:

- Handwritten rules and equations are too complex (face recognition).
- We cannot write the program ourselves.
- We cannot explain how (speech recognition).
- Need customized solutions to classify (spam or not).
- Rules are constantly changing (fraud detection).
- Traditional solutions cannot scale, as ML solutions are effective at handling large-scale problems.
- Develop systems that can automatically adapt and customize themselves to individual users (personalized news or mail filter).
- Discover new knowledge from large databases (Market Basket analysis), etc.

Moreover, it is hard to write programs and solve problems such as recognizing a face.

- We do not know what program to write because we do not fully comprehend how our brain does face recognition.
- Even if we have a good idea of how to do it, the program may be horrendously complicated.

Instead of writing a program by hand, we collect lots of examples that specify the correct output for a given input, also known as labeled dataset. A machine learning algorithm then takes these examples and produces a program that does the job.

- The program produced by the learning algorithm may look very different from a typical handwritten program. It may contain millions of numbers.
- If we do it right, the program works for new cases as well as the ones that we trained it on.

Finding hidden patterns and extracting key insights from input training data are the most important part of ML. By building predictive models using statistical techniques, ML enables us to dig beneath the surface and explore the patterns in data. Although understanding data and extracting patterns can be done manually, it will take lot of time, whereas ML algorithms can perform such tasks efficiently in less time.

1.10   Why Machine Learning Now?

This development is driven by a few underlying forces:

- With a reduction in the cost of sensors, the amount of data that can be generated and collected is increasing significantly, and over shorter periods of time. Data in 21st century is often compared to the role of oil in the 18th century [10], as an immensely and untapped valuable asset. Similar to oil driven prosperity [11], for those who can harness data's fundamental value, learn to extract and use it will reap huge rewards [3]. Nowadays, in digital economy, data is more valuable than ever. It is a key to success for various organizations from government to private companies. However, a large amount of data will be useless unless it is analyzed to find patterns hidden in the data.
- By making use of various algorithms, Machine learning can be used to make better business decisions [12]. For example, ML is used to forecast sales, predict stock price, identify risks, and fault detection.
- The cost of storing this data has reduced significantly.
- The cost of computing has come down significantly.
- Cloud has democratized compute for the masses and increased support from industries [13].

These ideas combine to create a world where we are generating more data, storing it cheaply, and running huge computations on it. This was not possible earlier, even though machine learning techniques and algorithms were fairly well known.

1.11   Classical Tasks for Machine Learning

In this section, we will discuss various tasks performed with machine learning:

- *Classification*:
  - Mining patterns that can classify future (new) data into known classes (e.g., whether to buy/sell a stock).
- *Clustering/grouping*:
  - Identify a set of similar groups in the data (e.g., customer segmentation, group the customers into different segments).
- *Prediction/regression*:
  - Predict the future value/behavior based on the past history (e.g., predict the future price of a stock).

- *Association rule mining*:

  - Mining any rule in the form of $X \rightarrow Y$, where X and Y are sets of data items, e.g., bread, butter $\rightarrow$ milk (transaction analysis, i.e., when people buy bread and butter, they will buy milk also).

- *Anomaly detection*:

  - Discover the outliers/unusual items in data (e.g., detect fraudulent activity on credit card).

1.12   Applications of Machine Learning

There are many applications of machine learning across many domains (e.g., pattern recognition, finance, natural language, computer vision, robotics, and manufacturing). Some applications include:

- Identify and filter spam messages from email
- Speech/handwriting recognition
- Object detection/recognition
- Predict the outcomes of elections
- Stock market analysis
- Search engines (e.g., Google)
- Credit card fraud detection
- Webpage clustering (e.g., Google News)
- Recommendation systems (e.g., Pandora, Amazon, Netflix)

1.12.1   Applications in Our Daily Life

Machine learning/artificial intelligence is everywhere. The possibility is that one is already using it, one way or another without realizing it, as shown in the following examples:

1. *Virtual personal assistant*: Siri, Alexa, and Google are some of the popular examples of virtual personal assistants.
2. *Predictions while commuting*: Traffic predictions, online transportation networks (e.g., when booking a cab, Uber app estimates the cost of a ride; while using these services, drivers minimize their detours in traffic).
3. *Video surveillance*: Monitoring multiple video cameras (e.g., alerts on intrusion detection).
4. *Social media services*: Suggesting people to befriend (e.g., by face recognition in uploaded pictures or matching preferences).

5. Email spam and malware filtering.
6. Product recommendation, online fraud detection.

1.13   ML Computing Needs

With the rise in machine learning and deep learning in every sector, the computing and storage needs for machine learning are growing. Understanding of the key technology requirements will help technologists, management, and data scientists tasked with realizing the benefits of machine learning to make intelligent decisions in their choice of hardware platforms. When trying to gain business value using ML, access to suitable hardware that supports all the complex functions is of utmost importance [4].

Machine learning entails building mathematical and probabilistic models, which require lot of computations. Among all processes of ML (i.e., pre-processing data, training ML algorithms, storing the trained model, and deployment of model), training the ML model is the most computationally intensive task. This process can be frustrating and time consuming if done without appropriate hardware. The first thing is to determine the nature of computational resources that a task requires [4].

If tasks are small and can fit in a complex sequential processing, one does not need a big system. One can even skip the GPUs altogether. A CPU such as i7-7500U can train an average of ~115 examples/s. So, unless one is planning to work on other ML areas or algorithms, a GPU is not necessary.

If a task is computationally intensive and has a manageable input dataset, then a reasonably powerful GPU would be a better choice. A laptop with a dedicated graphics card of high end should be sufficient. There are a few high end (and expectedly heavy) laptops like Nvidia GTX 1080 (8 GB VRAM), which can train an average of ~14k examples/s. In addition, one can build a PC with a reasonable CPU and a powerful GPU, but keep in mind that CPU must not bottleneck the GPU. For instance, an i7-7500U will work flawlessly with a GTX 1080 GPU. Requirement for best laptops for ML and recommendation for other laptops are given in [5], which are as follows:

RAM: A minimum of 16 GB is required, but 32 GB would be preferred.

CPU: Processors above **Intel Corei7 7th Generation** is advised as it is more powerful and delivers high performance.

GPU: This is the most important aspect as deep learning, a sub-field of machine learning. It requires artificial neural networks (ANN) to work, which are computationally demanding. Working on images or videos requires heavy amounts of matrix calculations. GPU enables parallel processing of these matrices. Without GPU, the process might take days or months. But with it, a laptop can perform the same for machine learning task in hours.

NVIDIA has started making GeForce 10 Series [6] for laptops. These are one of the best GPUs to work with. They have more advanced RTX 20 Series [7]. Other alternative is AMD Radeon [8].

Storage: A minimum of 1 TB HDD is required as datasets tend to get larger by the day. A system with a minimum of 256 GB SSD is advised. With less storage, one can opt for cloud storage options.

Operating System: Mostly People work with Linux, but Windows and MacOS can also run Virtual Linux Environment.

1.14 Machine Learning in the Cloud

As we noted in the previous section, if a minimum of 1 TB local storage is not available, then cloud storage option should be explored. Value of using elastic compute resources in cloud extends beyond storage, as enterprises collect data from multiple sources on a 24 × 7 basis. With the always up, and accessible from anywhere nature of the Cloud, it forms a natural repository to store all the incoming data. Furthermore, it can be immediately processed and labeled using ML algorithms. An example is to alert the sales team for new customer prospects.

Many benefits of conducting ML in the cloud are enumerated below [9]:

1. Cloud offers anytime, anywhere access.
2. Cloud's pay-per-use model is good for varying AI workloads.
3. Cloud makes it easy for enterprises to experiment with machine learning capabilities and scale up as projects go into production and demand increased resources.
4. Cloud makes intelligent capabilities accessible without requiring advanced skills in artificial intelligence or data science.
5. AWS, Microsoft Azure, and Google Cloud Platform offer many machine learning options that do not require deep knowledge of AI, machine learning theory, or a team of data scientists.

Specifically, Amazon offers SageMaker, which is a pre-installed framework with common ML algorithms. Google offers cloud ML engine that supports TensorFlow, and Microsoft offers Azure ML workbench and services that support Python-based frameworks, such as Tensorflow or PyTorch. The latter is also a desktop application that uses cloud-based ML services.

1.15 Tools Used in Machine Learning

There are several tools and languages being used in machine learning. The exact choice of the tool depends on your need and scale of operations. Here are the most commonly used tools in machine learning:

- Languages:
  - R
  - Python
  - SAS
  - Julia
  - Java
  - Javascript
  - Scala

- Databases:
  - SQL/NoSQL
  - Hadoop/Spark

- Visualization tools:
  - D3.js
  - Tableau
  - Seaborn

- Free and open-source software:
  - Scikit-learn [14]
  - Keras
  - Pytorch
  - Tensorflow
  - MXNet
  - Caffe
  - Weka

1.16 Points to Ponder

1. What is the difference between AI and ML?
2. What types of problems are suitable for unsupervised learning to solve?
3. What are the advantages and concerns of doing ML in a public cloud?

References

1. Samuel, A. L. (1959). Some studies in machine learning using the game of checkers. *IBM Journal of Research and Development., 44*, 206–226.
2. Mitchell, T. M. (1997). *Machine learning*. New York: McGraw-Hill International.
3. Bishop, C. M. (2006). *Pattern recognition and machine learning*. New York: Springer.
4. https://www.einfochips.com/blog/everything-you-need-to-know-about-hardware-require ments-for-machine-learning/.
5. https://www.edureka.co/blog/best-laptop-for-machine-learning/.

6. https://www.nvidia.com/en-us/geforce/products/.
7. https://www.nvidia.com/en-us/geforce/gaming-laptops/20-series/.
8. https://www.amd.com/en/graphics/radeon-rx-graphics.
9. https://cloudacademy.com/blog/what-are-the-benefits-of-machine-learning-in-the-cloud/.
10. https://www.wired.com/insights/2014/07/data-new-oil-digital-economy/.
11. https://www.economist.com/leaders/2017/05/06/the-worlds-most-valuable-resource-is-no-longer-oil-but-data.
12. https://en.wikipedia.org/wiki/Machine_learning.
13. https://software.intel.com/en-us/articles/intel-xeon-phi-delivers-competitive-performance-for-deep-learning-and-getting-better-fast?wapkw=deep-learning.
14. https://scikit-learn.org/stable/.

Chapter 2
Machine Learning Algorithms

The advances in science and technology improve the quality of our lives. Machine learning (ML) is an integral part of artificial intelligence (AI). It has spiked and is seen as playing a remarkable role in our lives. Machine learning is rapidly changing the world, with diverse applications and research pursued in industry and academia. It affects every part of our daily lives. For example, widely popular virtual personal assistants are being used for playing music or setting an alarm, while face detection or voice recognition applications are other useful examples of machine learning systems.

As in AI, algorithms are fundamental instruments used to solve ML problems. In other words, ML algorithms are at the core of every ML system. ML systems have the ability to learn and perform an expected task without human intervention. Figure 2.1 depicts how a prediction task is undertaken in ML [1]. The process shown clearly indicates that a desired prediction is realized through analysis of inputs and corresponding observations.

2.1 Why Choose Machine Learning?

One may wonder why choose machine learning? Simply put, machine learning makes managing complex tasks easier.

Depending on the desired functionality, choice of algorithms ranges from very basic to highly complex. It is important to make a wise selection of an algorithm to suit the nature of ML needs. Careful consideration and testing are needed before finalizing an algorithm. For example, linear regression works well for simple functions such as predicting the weather. In case, accuracy is important, then slightly higher-level functionalities such as neural networks are required. This concept is called "The Explainability–Accuracy Tradeoff." Fig. 2.2 explains this better.

It has been observed that no single algorithm works for all problems. For example, one cannot say that Random Forest is always better than Naïve Bayes, or

P. Gupta, N. K. Sehgal, *Introduction to Machine Learning in the Cloud with Python*, https://doi.org/10.1007/978-3-030-71270-9_2

Fig. 2.1 Machine learning
process

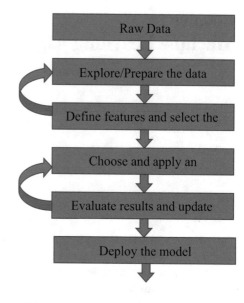

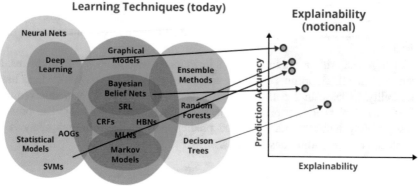

Fig. 2.2 The explainability–accuracy tradeoff

vice versa. That said, no one can deny that as practicing data scientists, we have to know the basics of common machine learning algorithms, which will help us to engage with the emerging problem domains that we come across in the real-world requirements.

A direct comparison of algorithms may be difficult as each one of it has its own features and functionality. There are many factors at work, such as the size and structure of the data sets, available computational time, urgency of the tasks, and the type of problems to be solved, e.g., classification or regression or prediction. Choosing an appropriate algorithm is a combination of the business needs, specifications, experimentations, available time, and resources. One should experiment with a variety of different algorithms for each specific class of problem and evaluate their performance and then select the best. While there are many algorithms that are present in the arsenal of machine learning, our focus will be on the most popular machine learning algorithms.

A machine learning model is a question/answer process that takes care of processing machine learning-related tasks. One can think of it as an algorithmic system that represents data when solving problems.

As discussed in Chap. 1, there are two major kinds of ML algorithms, i.e., supervised and unsupervised. In this chapter, we shall discuss both supervised and unsupervised ML algorithms to develop ML models and systems. These ML algorithms are well suited to develop both classification and prediction models.

2.2 Supervised Machine Learning Algorithms

In this method, to get the output for a new set of users' inputs, a model is trained to predict the results by using an old set of inputs and its known set of outputs. Output sought is also called target or a label. Basically, the system learns to predict from the past experience. Training data consists of a set of input and target pairs.

A data scientist trains the system by identifying the features and variables it should analyze. After training, these models compare the new results to the old ones, and update their weight's data accordingly to improve the future prediction accuracy.

Example: If there is a set of data where we have to classify gender into male and female, based on the earlier specifications like color, BMI, foot size, height, and weight given to the system, the model should be able to classify the gender.

There are two techniques in supervised machine learning. A technique to develop a model is chosen based on the type of data.

1. Regression
2. Classification

2.2.1 Regression

In regression, a numerical value or continuous variable is predicted based on the relationship between predictors (input variables) and output variable. An example would be predicting a house's price based on the current prices in the neighborhood, school district, total area, number of rooms, locality, and crime rate.

2.2.2 Classification

In classification, a categorical variable is predicted, i.e., input data can be categorized based on labels. For example, an email classification such as recognizing an email as spam or not a spam is a classification problem.

In summary, the regression technique is to be used when predictable data is quantified, and classification technique is to be used when predictable data is about predicting a label.

2.2.3 Machine Learning Algorithms: Supervised Learning

Below are some of the popular machine learning algorithms:

- Linear regression
- Logistic regression
- Support vector machines
- Naïve Bayes
- Decision Trees
- Random Forest
- Artificial neural networks
- K-nearest neighbors (KNN)

We shall discuss some of these algorithms in detail later in this chapter.

2.2.4 Machine Learning Algorithms: Unsupervised Learning

This method does not involve training the model based on old data, i.e., there is no "teacher" or "supervisor" to provide help to create a model with previous examples.

The system is not trained by providing a set of inputs and corresponding outputs. Instead, the model itself will learn and predict the output based on its own observations.

For example, consider a basket of apples and bananas which are not labeled/given any specifications this time. The model will only learn and organize them by comparing color, size, and shapes. This is achieved by observing specific features and similarities between the features.

We will discuss the techniques used in unsupervised learning as follows:

- Clustering
- Dimensionality reduction
- Anomaly detection

2.2.4.1 Clustering

Clustering is a method of dividing or grouping the data into clusters based on the observed similarities. Data is explored to make groups or subsets based on meaningful separations. For example, books in a library are put in a certain cluster based on the classification indices.

2.2.4.2 Dimension Reduction

If a dataset has too many features, it makes the process of segregation of data more complex. To solve complex scenarios, dimensionality reduction technique is used. The basic idea is to reduce the number of variables or features in a given dataset without losing its basic classification or prediction capability. Image classification can be considered as the best example where this technique is frequently used, by focusing on key attributes such as eyes, nose, and shape of face.

2.2.4.3 Anomaly Detection

Anomaly is an abnormality that does not fit with the rest of the pattern, e.g., bank fraud vs. regular transactions. Anomaly detection is the identification of such issues, using events or observations that raise suspicions by differing significantly from the majority of the data. Examples of the usage are identifying a structural defect in the manufacturing and medical problems such as detecting cancer.

2.2.5 Machine Learning Algorithms That Use Unsupervised Learning

Some of the common algorithms in unsupervised learning are:

- *K*-means clustering
- Hierarchical clustering
- DBSCAN
- Autoencoders
- Hebbian learning
- Deep belief nets
- Self-organizing map

We shall discuss some of these algorithms in detail later in this chapter.

2.3   Considerations in Choosing an Algorithm

Machine learning is both an art and a science. When we look at the machine learning algorithms, there is no one solution or one approach that fits all. Some problems are very specific and require a unique approach, e.g., if we look at a recommendation system, it is a very common type of machine learning algorithm and solves a very specific kind of problems. While some other problems are open and need a trial and error approach, these could be used in anomaly detection, or to build more general predictive models [2].

With advances in ML, there are several competing algorithms to choose from. Some algorithms are more practical than others in getting successful business results. There are several factors that can affect the decision to choose an ML algorithm that we will discuss in this section.

1. *Type of problem*: It is obvious that algorithms are designed to solve specific class of problems. So, it is important to know the type of problem we are dealing with and which kind of algorithm works better for the given problem. At a high level, ML algorithms can be classified into supervised and unsupervised. Supervised learning by itself can be categorized further into regression, classification, and anomaly detection.

2. *Understanding the nature of data*: For success in big data analytics, choosing the right dataset is paramount. The type of data plays a key role in deciding which algorithm to use. Some algorithms can work with smaller sample sets while others require a large number of samples. Some algorithms work well with certain type of data. For example, Naïve Bayes works well with categorical inputs but is not sensitive to missing data. Therefore, it is important to understand the nature of the data. Different algorithms may have different feature engineering requirements. Some have built-in feature engineering. Time spent on data extraction and feature engineering generally requires a large amount of time to be budgeted during the project development. If it is done properly, it is time well spent. There are various steps to be taken up while preparing the data before we feed it to ML algorithm. This process is known as pre-processing, which we will discuss later.

3. *Size of training set*: This is a major factor in our choice of algorithm. For a small training set, high bias/low variance classifiers (e.g., Naïve Bayes) have an advantage over low bias/high variance classifiers (e.g., KNN), since the later will over-fit. But low bias/high variance classifiers give better results as training set grows, i.e., they have lower asymptotic error, as high bias classifiers are not powerful enough to provide accurate models [3].

4. *Accuracy*: Depending on the application, the required accuracy will be different. Even an approximation is adequate sometimes. This may lead to huge reduction in processing time. In addition, approximate methods are very robust to over-fitting.

5. *Training time*: Various algorithms have different execution times. Training time is normally a function of the given dataset size and the desired accuracy.

6. *Linearity*: Lots of machine learning algorithms such as linear regression, logistic regression, and support vector machines make use of linearity. These assumptions are quite good for some problems. However, for some other problems, it may reduce the accuracy. In spite of these shortcomings, linear algorithms are very popular as a first line of attack. They tend to be algorithmically simpler and faster to train.

7. *Number of parameters*: Parameters affect the algorithm's behavior, such as error tolerance or number of iterations. Typically, algorithms with large number of parameters require trial and error to find a good combination. Even though having many parameters typically provides greater flexibility, training time and accuracy of the algorithm can sometimes be quite sensitive to getting just the right settings.

8. *Number of features*: The number of features in some datasets can be very large compared to the number of data points. This is often the case with genetics or textual data. The large number of features can bog down some learning algorithms, making training time infeasibly long. Some algorithms such as Support Vector Machines discussed later are particularly well suited to this case.

9. *Understanding system constraints*: It is important to know the system's data storage capacity. Depending on the storage capacity of the system, one might not be able to store gigabytes of classification/regression models or gigabytes of data to clusterize. This is the case, for instance, for embedded systems. Does the learning have to be fast? In some situations, training models quickly is necessary. Sometimes, one needs to rapidly update even on the fly.

10. *Find the available algorithms*: Once we have a better understanding of the problem and data, we can identify the algorithms that are applicable and practical to implement using the tools. Some of the factors affecting the choice of algorithm are:

- Whether the model meets the business goals?
- How much preprocessing the model needs?
- How accurate a model is?
- How explainable is the model?
- How fast is the model, i.e., how long does it take to build a model, and how long does the model take to make predictions?
- How scalable is the model?

An important criterion affecting choice of algorithm is model complexity. Generally speaking, a model is more complex if:

- It relies on more features to learn and predict (e.g., using two features vs. ten features to predict a target).
- It relies on more complex feature engineering (e.g., using polynomial terms, interactions, or principal components).
- It has more computational overhead (e.g., a single DT vs. a random forest of 100 trees).

Besides, the same machine learning algorithm can be made more complex based on the number of parameters or the choice of some hyper-parameters. In ML, a hyper-parameter is a parameter whose value is set before the learning process begins. For example,

- A regression model can have more features, or polynomial terms and interaction terms.
- A DT can have more or less depth.

Making the same algorithm more complex increases the chance of over-fitting, so one has to be careful with the complexity of the model.

2.4 What Are the Most Common and Popular Machine Learning Algorithms?

Following the general introduction to the machine learning algorithm types, let us now discuss some key machine leaning algorithms. Figure 2.3 shows an algorithm cheat sheet (which may be used as a rule of thumb) to choose a suitable algorithm. It has considered all the factors discussed earlier in making recommendation for choosing the right algorithm. It may not work for all situations and need to have a deeper understanding of these algorithms to pick the best algorithm for a given problem [13].

Sometimes more than one branch of algorithms will apply, and at other times, none of them will be a good match. It is important to remember these selections are intended to be rule-of-thumb recommendations. Some of the recommendations are not exact. In this section, we will cover a brief introduction to popular ML algorithms. Details such as mathematical or statistical analysis of these algorithms are beyond the scope of this book, readers are referred to [4–6].

2.4.1 Linear Regression

Linear regression is a very simple approach to supervised learning. Linear regression has been around for more than 200 years and has been an extensively studied technique. Regression models are highly valuable as one of the most common ways to make inferences and predictions. Regression is a statistical method to establish a relationship between a dependent variable and a set of independent variables. For instance, we want to find the relation between the price of the house and income level, crime rate or school district, etc. Linear regression is a basic predictive analytics technique that uses historical data to predict an output variable. It is popular for predictive modeling because it is easily understood and can be explained using plain English. Linear regression models have been to predict

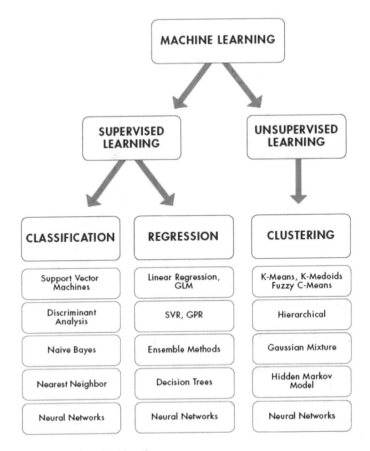

Fig. 2.3 Machine learning algorithm chart

growth, stock price, employee performance, blood pressure levels, and more. Understanding how to implement linear regression models can help to unearth stories in data to solve important problems.

It was developed in the field of statistics. The relationship is established between the independent and dependent variables by fitting the best line/hyper plane (for two-dimensional/multi-dimensional instances). The algorithm exposes the impact on the dependent variable by changing the independent variable. The independent variables are referred to as explanatory variables, as they explain the factors that impact the dependent variables. Dependent variable is often referred to as the predictor. Basically, the regression technique finds out a linear relationship between input and output, hence, the name linear regression. If we plot the independent variable on the x-axis and dependent variable on the y-axis, linear regression gives a straight line which best fits the data point, as shown in Fig. 2.4.

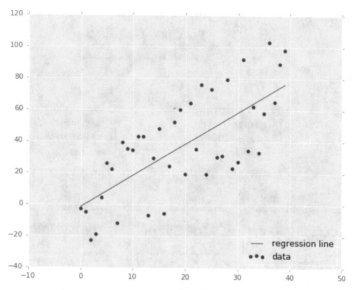

Fig. 2.4 An example of linear regression

2.4.2 Two Types of Linear Regression

2.4.2.1 Simple Linear Regression

The easiest model is the simple linear regression model with one input and one output. Based on the value of the single explanatory variable, the value of the response variable changes. The equation of a simple linear regression model to calculate the value of the dependent variable, y based on the predictor x, is as follows:

$$y_i = \beta_0 + \beta_i x + \varepsilon$$

where the value of y_i is calculated with input variable x_i for every ith observation. β's are known as regression coefficients. The ith value of x has ε_i as its error in the measurement. These coefficients determine the goodness of predictions of our model. The goal is to find statistically significant values of the parameters β's that minimize the difference between y and y_i (predicted value). If we are able to determine the optimum values of these parameters, then we will have the line of best fit that can be used to predict the value of output given a new value of input. The best fit line is the one for which the total prediction error is minimum. This method is known as ordinary least squares (OLS). The OLS technique tries to reduce the sum of squared error defined in the following equation by finding the best possible values of regression coefficients. OLS uses squared error that has nice mathematical properties, thereby making it easier to differentiate and compute gradient descent. The residual is calculated as:

$$e_i = y_i - \widehat{y}_i$$

and squared error is defined as:

$$J(\beta) = \frac{1}{n} \sum_{i=1}^{n} e_i^2$$

2.4.2.2   Multiple Linear Regression

The value of response variable is dependent upon more than one explanatory variables. Therefore, the simple linear regression model cannot be utilized, as there is a need for undertaking multiple linear regression for analyzing the predictor variables. Linear regression involving multiple variables is called "multiple linear regression." The steps to perform multiple linear regression are almost similar to that of a simple linear regression. The difference lies in the evaluation. One can use it to find out which factor has the highest impact on the predicted output and how different variables relate to each other. The equation of multiple linear regression is as follows:

$$y = \beta_0 + \beta_1 x_1 + \beta_2 x_2 + \cdots + \beta_p + \varepsilon$$

The x's are explanatory variables and determine y.

Some good rules of thumb when using this technique are to remove any variables that are correlated and to remove noise/outliers from the input data.

2.4.2.3   Assumptions of Linear Regression

Using linear regression requires that the model should conform to the following assumptions:

- The regression model is linear in parameters (which are coefficients and the error term).
- Linear regression requires residuals should be normally distributed. If the maximum likelihood (not OLS) is used to compute the estimates, then this implies that the dependent and independent variables are also normally distributed.
- The mean of residuals is zero. Error term actually refers to the variance present in the response variable that the independent variables failed to explain. The model is said to be unbiased if the mean of the error variable is zero.
- Variance of the residuals is constant, i.e., residuals are evenly distributed around mean or residuals are approximately equal for all predicted dependent variable

values. This condition is known as *homoscedasticity*. If the variance changes, it is referred as *heteroscedasticity*.

- Independent variables should not be perfectly correlated (i.e., no multi-collinearity). Perfect correlation between two variables suggests that they contain same information in them. In other words, both the variables are different forms of the same variable. If variables are correlated, it becomes extremely difficult for the model to determine the true effects of independent variables on dependent variable.
- Residuals should not be correlated with each other. This problem is also known as autocorrelation. This is applicable especially for time series data. Autocorrelation is the correlation of a time series with lags of itself. When the residuals are correlated, it means that the current value is dependent on the previous values and that there is a definite unexplained pattern in a dependent variable that shows up in the disturbances.
- Residuals should not be correlated with the independent variables. If residuals are correlated with the independent variable, one can use the independent variables to predict the error. This correlation between error terms and independent variables is known as endogeneity. When this kind of correlation occurs, a model may attribute the variance present in error to the independent variable, which in turn produces incorrect estimates.

2.4.2.4 Advantages

- Linear regression is one of the most interpretable machine learning algorithms. It is also easy to explain [14].
- It is easy to use, as it requires minimal tuning.
- It is the most widely used machine learning technique.

2.4.2.5 Disadvantages

- Model makes strong assumptions about the data.
- It works on only numeric features, so categorical data requires extra-processing.
- It does not do well with missing values and in the presence of outliers.

2.4.2.6 Sample Python Code for Linear Regression

Documentation of sklearn Linear Regression:

https://scikit-learn.org/stable/modules/generated/sklearn.
linear_model.LinearRegression.html

importing required libraries

```
import pandas as pd
from sklearn.linear_model import LinearRegression
from sklearn.metrics import metrics
from sklearn,model_selection import train_test_split

# read the train and test dataset
train_data = pd.read_csv('train.csv')
test_data = pd.read_csv('test.csv')

### Split the data into training and testing

train_x,test_x,train_y,test_y = tarin_test_split(X,y,test_size =
0,2, random_state = 1)
print(train_data.head())
Create the object of the Linear Regression model

Linear_model = LinearRegression()

# fit the model with the training data
linear_model.fit(train_x,train_y)

# coefficients of the trained model
print('\nCoefficient of model :', linear_model.coef_)

# intercept of the model
print('\nIntercept of model',linear_model.intercept_)

# How good is the fit
plt.scatter(train_x, train_y)
plt.plot(train_x, linear_model.predict(train_x), color = 'red')

# predict the target on the test dataset
predict_test = model.predict(test_x)

# Test data and predicted data
plt.scatter(test_x, test_y)
plt.plot(test_x, linear_model.predict(test_x), color = 'red')

# Evaluating the Algorithm
print('Mean Absolute Error:', metrics.mean_absolute_error(y_test,
y_pred))
print('Mean Squared Error:', metrics.mean_squared_error(y_test,
y_pred))
print('Root Mean Squared Error:', np.sqrt(metrics.mean_squared_error
(y_test, y_pred)))
```

2.4.3 K-Nearest Neighbors (KNN)

KNN is a very simple and effective machine learning algorithm. It is a non-parametric, lazy-learning algorithm, which means that there is no explicit training phase before classification. KNN algorithm uses the entire dataset as the training set, i.e., it does not split the dataset into training and testing sets. KNN algorithm assumes that similar entities exist in close proximity. In other words, similar entities have features that are close to each other. Figure 2.5 shows similar data points that are close to each other. KNN algorithm hinges on this assumption being true enough for the algorithm to be useful. KNN can be used to predict loan approval, calculate credit ratings, speech recognition, handwriting detection, image recognition, intrusion detection, etc. KNNs are used in real-life scenarios where non-parametric algorithms are required. These algorithms do not make any assumptions about how the data is distributed. So, not having to worry about the distribution is a big advantage. It means that KNN can be applied to a variety of datasets.

The main purpose of KNN algorithm is to classify the data into several classes to predict the class of a new data point. The K-nearest neighbors algorithm estimates how likely a data point is to be a member of one group or another. It essentially looks at the data points around a single data point to determine which group it belongs to. For example, if one point is on a grid and the algorithm is trying to determine the group that this data point is in (Group A or Group B, for example), it would look at the data points close to it to see the group majority of the points are in. In KNN algorithm, the predictions are made for a new data set by searching through the entire training set for the K similar instances, the neighbors, and summarizing the output variable for those K instances.

In the classification setting, the KNN algorithm essentially computes majority vote from the K most similar instances to a given "unseen" observation. Similarity is

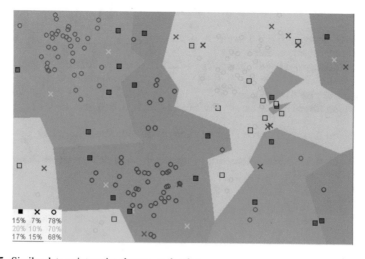

Fig. 2.5 Similar data points exist close to each other

defined according to a distance metric between two data points. A popular choice is the Euclidean distance given by

$$d(x, x') = \sqrt{(x_1 - x'_1)^2 + \cdots + (x_n - x'_n)^2}$$

where x and x' are two vectors, while d is the distance between them. However, other measures of distance can be more suitable for a given setting and include the Manhattan, Chebyshev, and Hamming distance [4, 5].

Given a positive integer K, a new observation x, and a similarity metric d, KNN classifier performs the following two steps:

- It runs through the whole dataset, computing d between x and each training observation. We will call the K points in the training data that are closest to x the set A. Note that K is usually odd to prevent tie situation.
- It then estimates the conditional probability for each class, that is, the fraction of points in A with that given class label. Note $I(x)$ is the indicator function which evaluates to 1 when the argument x is true and otherwise false.

$$p(y = j | X = x) = \frac{1}{K} \sum_{i \in A} I\left(y^{(i)} = j\right)$$

Finally, x gets assigned to the class with the largest probability.

KNN is memory-intensive and performs poorly for high-dimensional data and requires a meaningful distance function to calculate similarity.

2.4.3.1 Assumptions

- KNN assumes that the data is in some *identifiable feature space*. More precisely, the data points are in a metric space. The data can be scalars or could possibly be from multidimensional vector space. Since the points are in feature space, they have a notion of distance. This need not necessarily be Euclidean distance although it is the one commonly used.
- Each of the training data consists of a set of vectors and class label associated with each vector. In the simplest case, it will be either + or − (for positive or negative classes). But KNN works equally well with arbitrary number of classes.

We work with a number "K". This number decides how many neighbors (where neighbors are defined based on the distance metric) influence the classification. This is usually an odd number. If $K = 1$, then the algorithm is simply called the nearest neighbor algorithm.

2.4.3.2 How Does KNN Algorithm Works?

1. Select the K where K is the number of neighbors.
2. Compute the distance between the test point and each point.
3. Sort the distances in ascending (or descending) order.
4. Use the sorted distances to select the K-nearest neighbors.
5. Use majority rule (for classification) or averaging (for regression) to assign the class or value, respectively.

2.4.3.3 Advantages

- Simple to understand and easy to implement
- Zero to very small training time
- Works just as easily with multiclass datasets

One of the obvious drawbacks of the KNN algorithm is that it has a computationally expensive testing phase. Furthermore, KNN may sometimes suffer from a skewed class distribution.

Sample Python Code for KNN Algorithm

Documentation of sklearn K-Neighbors Classifier:

https://scikit-learn.org/stable/modules/generated/sklearn.
neighbors.KNeighborsClassifier.html

```
from sklearn.neighbors import KNeighborsClassifier
from sklearn.metrics import accuracy_score
```

Create the object of the K-Nearest Neighbor model with 3 neighbors

```
knn_model = KNeighborsClassifier(n_neighbors =3)

# fit the model with the training data
knn_model.fit(train_x,train_y)

# Number of Neighbors used to predict the target
print('\nThe number of neighbors used to predict the target : ',
knn_model.n_neighbors)

# predict the target on the test dataset
predict_test = knn_model.predict(test_x)
print('Target on test data',predict_test)
```

```
# Evaluating the Algorithm
accuracy_test = accuracy_score(test_y,predict_test)
print('accuracy_score on test dataset : ', accuracy_test)

print(confusion_matrix(test_y,predict_test))
```

2.4.4   Logistic Regression

Logistic regression is a classic predictive modeling technique and still remains a popular choice for modeling binary categorical variables [15]. Logistic regression is the classification counterpart of linear regression. It is also one of the most popular supervised machine learning algorithms. It is used to predict the categorical dependent variable using a given set of independent variables. Logistic regression is an efficient and powerful way to analyze the effects of a group of independent variables with binary outcomes by quantifying each independent variable's unique contribution to predict the output of a categorical dependent variable. Therefore, the outcomes must be categorical or should have discrete values. Each can be either yes or no, 0 or 1, true or false, etc., but instead of giving the exact value as 0 or 1, it gives the probabilistic values which lie between 0 and 1. This method can be used in spam detection, credit card fraud, predicting whether a given mass of tissue is benign or malignant, etc.

The name of this algorithm can be a little confusing in the sense that logistic regression algorithm is used for classification tasks. In fact, logistic regression is much similar to the linear regression except that how it is used. Linear regression is used for regression problems, whereas logistic regression is used for solving the classification problems. The name "Regression" implies that a linear model is to fit into the linear space. Predictions are mapped to be between 0 and 1 through the logistic function shown in Fig. 2.6 to a linear combination of features [16], which

Fig. 2.6 Logistic function

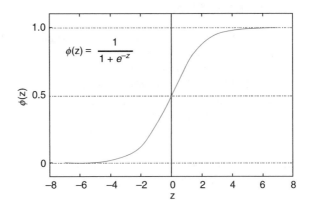

$$\phi(z) = \frac{1}{1 + e^{-z}}$$

means that predictions can be interpreted as class probabilities. The odds or probabilities that describe the outcome of a single trail are modeled as a function of
explanatory variables. As the models themselves are still "linear," they work well
when the classes are linearly separable (i.e., they can be separated by a single
decision surface). In logistic regression, instead of fitting a regression line, we fit
an "S"-shaped logistic function, shown in Fig. 2.6, which predicts two maximum
value (0 or 1). The curve from the logistic function indicates the likelihood of
something such as whether the email is spam or not.

Using the components of linear regression reflected in logit scale, logistic regression iteratively identifies the strongest linear combination of variables with the
highest probability of detecting the observed outcome. The output is generated by
a logarithmic transformation of the input values, defined by the logistic function f
$(x) = 1/(1 + e^{-x})$. A threshold is then applied to force this probability into a binary
classification. The goal of logistic regression is used to train the model to find the
values of coefficients such that it will minimize the error between the predicted
outcome and the actual outcome. These coefficients are estimated using maximum
likelihood estimation.

2.4.4.1 Types of Logistic Regression

On the basis of categories, logistic regression can be classified into three types:

- *Binomial*: In binomial logistic regression, there can be only two possible types of
  the dependent variables, such as 0 or 1, yes or no.
- *Multinomial*: In multinomial logistic regression, there can be three or more types
  of dependent (categorical) variables. Multinomial logistic regression can be used
  to distinguish cats from dogs or sheep.
- *Ordinal*: In ordinal logistic regression, there can be three or more possible ordered
  types of dependent variables, such as small, medium, or large.

2.4.4.2 Assumptions

Logistic regression does not make many of the key assumptions of linear regression
and general linear models based on ordinary least squares algorithms. In particular,
regarding linearity, normality, homoscedasticity, and measurement level are done
away with. First, it does not need a linear relationship between the dependent and
independent variables. The primary reason why logistic regression can handle all
sorts of relationships is because it applies a nonlinear log transformation to the
predicted odds ratio. Second, the independent variables do not need to be multivariate normal—although multivariate normality yields a more stable solution. Also, the
error terms (the residuals) do not need to be multivariate normally distributed.
Logistic regression does not need variances to be heteroscedastic for each level of
the independent variables. Finally, it can handle ordinal and nominal data as

independent variables. The independent variables do not need to be metric (interval or ratio scaled). However, some other assumptions still apply which are described below:

- The outcome is a binary or dichotomous variable like true or false, 1 or 0. Reducing an ordinal or even metric variable to dichotomous level loses a lot of information, which makes this test inferior compared to ordinal logistic regression in these cases.
- Logistic regression assumes linearity of independent variables and log odds. While it does not require the dependent and independent variables to be related linearly, it requires that the independent variables are linearly related to the log odds. In other words, there is a linear relationship between the logit of the outcome and each predictor variable.
- There are no influential values (extreme values or outliers).
- The model should be fitted correctly. Neither over-fitting nor under-fitting should occur. That is only the meaningful variables should be included. A good approach to ensure this is to use a stepwise method to estimate the logistic regression.
- It requires quite a large sample size. Because maximum likelihood estimates are less powerful than ordinary least squares (e.g., simple linear regression, multiple linear regression). It has been observed that OLS needs 5 cases per independent variable in the analysis, ML needs at least 10 cases per independent variable, and some statisticians recommend at least 30 cases for each parameter to be estimated.
- There is no high intercorrelations (i.e., multicollinearity) among the predictors.

2.4.4.3   Advantages

- It is easier to inspect and less complex.
- It is a robust algorithm as the independent variables need not have equal variance or normal distribution.
- The algorithm does not assume a linear relationship between the dependent and independent variables and hence can also handle nonlinear effects.
- The algorithm can be regularized to avoid over-fitting.
- The model can be easily updated with new data using stochastic gradient descent.

2.4.4.4   Disadvantages

- Logistic regression tends to underperform when there are multiple or nonlinear decision boundaries.
- Model is not flexible enough to naturally capture complex relationships.
- Logistic models tend to over-fit the data when the data is sparse and of high dimension. It requires more data to achieve stability and meaningful results.
- It is not robust to outliers and missing values.

Sample Python Code for Implementing Logistic Regression

```
Documentation of sklearn LogisticRegression:
https://scikit-learn.org/stable/modules/generated/sklearn.
linear_model.LogisticRegression.html

# importing required libraries
import pandas as pd
from sklearn.linear_model import LogisticRegression
from sklearn.metrics import accuracy_score

# read the train and test dataset
train_data = pd.read_csv('train-data.csv')
test_data = pd.read_csv('test-data.csv')

Create the object of the Logistic Regression model ,

model = LogisticRegression()

# fit the model with the training data
model.fit(train_x,train_y)

# coefficients of the trained model
print('Coefficient of model :', model.coef_)

# intercept of the model
# print('Intercept of model',model.intercept_)

# predict the target on the test dataset
predict_test = model.predict(test_x)
print('Target on test data',predict_test)

# Accuracy Score on test dataset
accuracy_test = accuracy_score(test_y,predict_test)
print('accuracy_score on test dataset : ', accuracy_test)
print(confusion_matrix(test_y,predict_test))
```

2.4.5   Naïve Bayes Classifier Algorithm

It is called naïve because it assumes that the occurrence of a certain feature is independent of the occurrence of other features. Such as if the fruit is to be identified on the basis of color, shape, and taste, then a yellow, spherical, and sweet fruit is recognized as an orange. Hence, each feature individually contributes to identify that it is an orange without depending on each other. It is called Bayes' because it depends on the principle of Bayes' Theorem.

Bayes' Theorem:

- Bayes' theorem is also known as Bayes' rule or Bayes' law, which is used to determine the probability of a hypothesis with prior knowledge. It depends on the conditional probability.
- The formula for Bayes' theorem is given as:

$$P(A|B) = \frac{P(B|A)P(A)}{P(B)}$$

  where
- $P(A|B)$ is posterior probability. Probability of hypothesis A on the observed event B.
- $P(B|A)$ is likelihood probability. Probability of the evidence given that the probability of a hypothesis is true.
- $P(A)$ is prior probability. Probability of hypothesis before observing the evidence.
- $P(B)$ is marginal probability. Probability of evidence.

Using the Bayes' theorem, it is possible to build a learning system that predicts the probability of the response variable belonging to some class, given a new set of attributes.

An NB model comprises two types of probabilities that can be calculated directly from the training data:

1. The probability of each class
2. The conditional probability for each class given each input value

A Naïve Bayes model multiplies several different calculated probabilities together to identify the posterior probability for each class and select the class with the highest probability. This is called the maximum a posteriori probability. Conditional independence of the features reduces the complexity of model.

Naïve Bayes *classifiers* are linear classifiers that are known for being simple yet very efficient. Naïve Bayes (NB) classifier is among the most popular learning method grouped by similarities that works on the popular Bayes Theorem of probability with an assumption of independence between predictors—to build ML models. It is a probabilistic classifier, which predicts on the basis of conditional probability for its likely classification. Along with simplicity, Naïve Bayes is known to outperform even highly sophisticated classification methods. The method has a strong assumption of independent input variables which is not always realistic for real data. Nevertheless, the technique is very effective on large range of complex problems. If the NB conditional independence assumption actually holds, a Naïve Bayes classifier will converge quicker than the discriminative models such as logistic regression. Also, one needs less data to train the model. Even if the NB assumption of independence of features does not hold, it still often does a good job in practice. NB algorithm finds usage in text classification, spam filtering, sentiment analysis, recommendation system, etc. However, strong violations of the independent assumptions and nonlinear classification problems may lead to poor performances of Naïve Bayes classifiers.

2.4.5.1   Additive Smoothing

During testing time, if we come across a feature that we did not come across during training time, then the individual conditional probability of that particular feature will become zero, thus making class-conditional probabilities equal to zero. So, we have to modify the formula for calculating individual conditional probability. In order to avoid the problem of zero probabilities, an additional smoothing term can be added to the multinomial Bayes model.

2.4.5.2   Types of Naïve Bayes Model

- *Gaussian*: The Gaussian model assumes that features follow a normal distribution. This means if predictors take continuous values instead of discrete, then the model assumes that these values are sampled from the Gaussian distribution.
- *Multinomial*: The multinomial Naïve Bayes classifier is used when the data is multinomial distributed. It is primarily used for document classification problems. It means a particular document belongs to which category such as politics, sports, and entertainment. The features/predictors used by the classifier are the frequency of the words present in the document.
- *Bernoulli*: The Bernoulli classifier s works similar to the multinomial classifier, but the predictor variables are the independent Boolean variables, for example, if a particular word is present or not in a document.

2.4.5.3   Assumptions

- The fundamental Naïve Bayes assumption is that each feature makes an independent and equal (i.e., identical) contribution to the outcome.

2.4.5.4   How Naïve Bayes Algorithm Works?

Following steps are followed while working with this algorithm:

1. Calculate prior probability for the given class labels.
2. Calculate conditional probability with each attribute for each class.
3. Multiply same class conditional probability.
4. Multiply prior probability with probability obtained in step 3.
5. Determine which class has a higher probability. Higher probability class belongs to the given input set.

2.4.5.5   Advantages

- NB models performs well in practice even when conditional independence assumption rarely holds true.
- It is easy to implement and can scale with data.
- It is a good choice when CPU and memory resources are a limiting factor.
- It performs well when the input variables are categorical.
- It is easier to predict class of the test data set.

2.4.5.6   Disadvantages

- NB assumes that all features are independent or unrelated, so it cannot learn the relationship between features.

Sample Python Code for Naïve Bayes Model

```
Documentation of sklearn GaussianNB:

https://scikit-learn.org/stable/modules/generated/sklearn.
naive_bayes.GaussianNB.html

from sklearn.naive_bayes import GaussianNB
from sklearn.metrics import accuracy_score

Create the object of the Naive Bayes model

model = GaussianNB()

# fit the model with the training data
model.fit(train_x,train_y)

# predict the target on the test dataset
predict_test = model.predict(test_x)
print('Target on test data',predict_test)

# Accuracy Score on test dataset
accuracy_test = accuracy_score(test_y,predict_test)
print('accuracy_score on test dataset : ', accuracy_test)
```

2.4.6   Support Vector Machine Algorithm

Support vector machines (SVM) are supervised machine learning algorithms which can be used for classification and regression problems as support vector classification (SVC) and support vector regression (SVR) while they can be used for regression. SVM is mostly used for classification. SVM algorithm can be used for face detection, image classification, text categorization, etc. They were extremely popular around the time when developed in the 1990s and continue to be the go-to methods for high-performing algorithm with some tuning.

While it sounds simple, not all datasets are linearly separable. In fact, in the real world, almost all the data is randomly distributed, which makes it hard to separate different classes linearly. SVM uses a kernel trick to solve such type of problems. Kernel trick performs some data transformations to figure out an optimal boundary to separate data based on the labels or outputs defined. Kernel trick is a method of using linear classifier to solve nonlinear problems. It entails transforming seemingly linearly inseparable data. The kernel function maps the original nonlinear observations into a higher-dimensional space in which they become separable as shown in Fig. 2.7. In this figure, we find a way to map the data from two-dimensional to three-dimensional space and are able to find a decision boundary that clearly divides between different classes. Kernel trick allows us to operate in the original feature space without computing the coordinates of the data in a higher dimensional [7]. There are many kernel functions e.g., Gaussian/RBF kernel polynomial kernel, and sigmoid. Care has to be taken in choosing kernel function to avoid over-fit the model. Thus, choosing the right kernel function (including the right parameters) and regularization (discussed in other part of the book) are of great importance. The most used type of kernel function is RBF, because it has localized and finite response [8].

SVM is based on the idea of finding a hyper-plane that best separates the features into different domains. SVM essentially calculates distance between two observations. The objective of the SVM is to find a hyper-plane in N-dimensional space (N being the number of features) that distinctly classifies the data. In other words, the idea behind SVM is to find decision planes that define decision boundaries. A decision plane is one that separates a set of objects having different class memberships. As there are many such linear hyper-planes, SVM algorithm tries to maximize

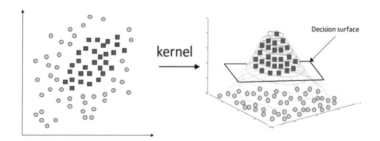

Fig. 2.7 Kernel method

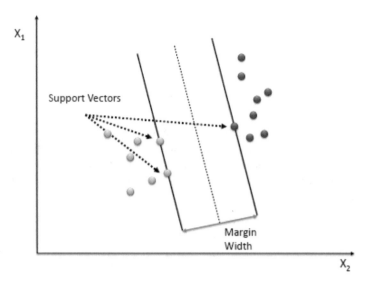

Fig. 2.8 Support vector machines

the distance between the various classes that are involved, and this is referred to as margin maximization. If the line that maximizes the distance between the classes is identified, then the probability to generalize well to new data is enhanced. An example of binary classes is shown in Fig. 2.8. The points closest to the hyper-plane are called support vectors (SV), and the distance of the vectors from the hyper-plane is called margin. The basic intuition to develop here is that the farther the support vector points are from the hyper-plane, higher is the probability of correctly classifying points in their respective regions or classes. SV points are very critical in determining the hyper-plane because if the position of vectors changes, then the position of the hyper-plane is altered. Technically, this hyper-plane can also be referred as maximizing margin hyper-plane. If an SVM is given a data point closer to the classification boundary than the support vectors, then SVM declares that data point to be too close for accurate classification. This defines a "no-man's land" for all points within the margin of the classification boundary. Since the support vectors are the data points closest to this "no-man's land" without being in it, intuitively they are also the points most likely to be misclassified.

Kernel tricks are used to map a nonlinearly separable function into a higher dimension linearly separable function. A support vector machine training algorithm finds the classifier represented by the normal vector and bias of the hyper-plane. This hyper-plane (boundary) separates different classes by as wide a margin as possible.

An SVM finds the classifier represented by the normal vector and bias.

2.4.6.1 Assumptions

SVMs can be defined as linear classifiers under the following two assumptions:

- The margin should be as large as possible.
- The support vectors are the most useful data points because they are the ones most likely to be incorrectly classified.

The second assumption leads to a desirable property of SVMs. After training, the SVM can throw away all other data points and just perform classification using the support vectors, for example, in the above Fig. 2.7, we need only three points. This means that once classification is done, an SVM can predict the class of a data point very efficiently.

2.4.6.2 Types of SVM

SVM can be of two types:

- *Linear SVM*: Linear SVM is used for linearly separable data, which means if a dataset can be classified into two classes by using a single straight line, then such data is termed as linearly separable data, and the classifier used is called a linear SVM classifier.
- *Nonlinear SVM*: A nonlinear SVM is used to separate dataset that cannot be classified by using a straight line. Such data is termed as nonlinear data, and the classifier is called as nonlinear SVM classifier.

2.4.6.3 Advantages

- It works well on a smaller and cleaner dataset.
- It is more efficient as it uses a subset of training points.
- It can model nonlinear boundaries, and there are many kernels to choose from.
- It is robust against over-fitting, especially high-dimensional space.

2.4.6.4 Disadvantages

- It is not suited for larger datasets because of training time.
- It is less effective on noisier dataset with overlapping classes.
- It is memory intensive.
- It is trickier to tune due to the importance of picking the right kernel.
- It does not scale well to large dataset.

Sample Python Code for SVM

```
Documentation of sklearn Support Vector Classifier:

https://scikit-learn.org/stable/modules/generated/sklearn.svm.
SVC.html

from sklearn.svm import SVC
from sklearn.metrics import accuracy_score

Create the object of the Support Vector Classifier model

model = SVC()

# fit the model with the training data
model.fit(train_x, train_y)

# predict the target on the test dataset
predict_test = model.predict(test_x)
print('Target on test data', predict_test)

# Accuracy Score on test dataset
accuracy_test = accuracy_score(test_y, predict_test)
print('accuracy_score on test dataset : ', accuracy_test)
```

2.4.7 Decision Trees

Decision Trees (DTs) belong to supervised machine learning algorithms. The tree can be explained by two entities, namely decision nodes and leaves. The leaves are the final outcomes, and each node (decision nodes or internal nodes) within the tree represents a test on specific feature.

These methods are used for regression as well as classification problems. Using the DT with a given set of inputs, one can map the various outcomes that are a result of the consequences or decisions. These trees are used to provide graphical outputs to the user based on several independent variables. DT is capable of handling heterogeneous as well as missing data. DT algorithms are further capable of producing comprehensible rules. Classification can be performed without many computations. DTs are used in marketing and sales, reducing churn rate, anomaly and fraud detection, medical diagnosis, etc.

A DT is a tree-like graph. At each node, an edge is generated based on some specific characteristics of one of the attributes. We may well state that nodes represent a place where we pick an attribute and ask a question, and edges represent the answers to the question. Finally, the leaves represent actual output or class labels.

They are used in nonlinear decision-making with a simple linear decision surface. A DT is largely used for non-parametric machine learning modeling for regression and classification problems. To find solutions, a DT makes sequential, hierarchical decision about the outcome variable based on the predictor data. So, what does all that mean?

Hierarchical means that the model is defined by a series of questions that lead to a class label or a value when applied to any observation. Once set up, the model acts like a protocol in a series of "if this occurs then that follows" conditions that produce a specific result from the input data.

A non-parametric method means that there are no underlying assumptions about the distribution of the errors or the data. It basically means that the model is constructed based on the observed data.

The understanding level of DT algorithm is so easily compared with other classification algorithms. DT algorithm can be used for both regression and classi-fication. DT method is capable of handling heterogeneous as well as missing data. Trees are further capable of producing understandable rules. DTs often mimic the human-level thinking, so it is simpler to understand the data and make some good interpretations. DTs make it easy to see the logic for the data to interpret (unlike SVM, ANN, etc.)

This algorithm is known as Hunt's algorithm, which is both greedy and recursive. Greedy means that at each step, it evaluates to locally maximize (minimize) an objective function. Similarly, recursive means that it splits the larger question into smaller questions and resolves them using the same argument. The decision to split at each node is made according to the metric called **purity**. A node is 100% impure when a node is split evenly 50/50 and 100% pure when all of its data belongs to a single class. In order to optimize our model, we need to reach maximum purity and avoid impurity.

DT may use multiple criteria to decide to split a node in two or more sub-nodes. With each split, purity of the node is enhanced with respect to the target variable. DT split the nodes on all the available attributes and select the split, which results in most homogeneous sub nodes. Given a data table that contains attributes and class of the attributes, we can measure homogeneity (or heterogeneity) of the table based on the classes. We say a table is pure or homogeneous if it contains only a single class. If a data table contains several classes, then we say that the table is impure or heteroge-neous. There are several indices to measure the degree of impurity quantitatively. Some well-known indices to measure the degree of impurity are entropy, Gini index, and classification error.

The popular measures used are:

- Information gain
- Gini index

2.4.7.1   Information Gain

Information gain is the measurement of changes in entropy after the segmentation of a dataset based on an attribute. It calculates how much information a feature provides about a class. According to the value of information gain, we split the node and build the DT. A DT algorithm always tries to maximize the value of information gain, and a node/attribute having the highest information gain is split first. It can be calculated using the formula given below:

Information gain = Entropy (S) − (weighted avg) * Entropy (for each attribute)

Entropy is a metric to measure the impurity in a given feature. It specifies randomness in data. Entropy can be calculated as:

$$\text{Entropy}(s) = \sum_j - p_j \log_2 p_j$$

where p_j is the probability of class j and s is the total number of data points.

2.4.7.2   Gini Index

Gini index is a measure of impurity or purity used while creating a decision boundary. An attribute with the low Gini index should be preferred as compared to the high Gini index. Gini index can be calculated using the formula below:

$$\text{GiniIndex} = 1 - \sum_j p_j^2$$

These measures will calculate values for every attribute. The values are sorted, and attributes are placed in the tree by following the order, i.e., the attribute with a high value (in case of information gain) is placed at the root. While using information gain as an algorithm, we assume attributes to be categorical, and for Gini index, attributes are assumed to be continuous.

2.4.7.3   Decision Tree Terminology

- *Root node*: Root node is from where the DT starts. It represents the entire dataset, which further gets divided into two or more homogeneous sets.
- *Leaf node*: Leaf node is the final output node, and the tree cannot be segregated further after getting a leaf node.
- *Splitting*: Splitting is the process of dividing the decision node/root node into sub-nodes according to the given conditions.
- *Branch/sub-tree*: A tree formed by splitting the tree.

- *Pruning*: Pruning is the process of removing the unwanted branches from the tree.
- *Parent/child node*: The root node of the tree is called the parent node, and other nodes are called the child nodes.

2.4.7.4   Assumptions

Below are some of the assumptions we make while using DT:

- At the beginning, the whole training set is considered as the root.
- Feature values are preferred to be categorical. If the values are continuous, then they are discretized prior to building the model.
- Records are distributed recursively on the basis of attribute values.
- Order to placing attributes as root or internal node of the tree is done by using some statistical approaches, which are mentioned above.

2.4.7.5   How Does the Decision Tree Classifier Work?

The complete process can be better understood using the algorithm given below:

1. The tree is constructed in a top-down recursive manner.
2. At start, all the training examples are at the root.
3. Examples are partitioned recursively based on selected attributes.
4. Attributes are selected on the basis of an impurity function (e.g., information gain).

Condition for stopping partitioning:

- All examples for a given node belong to the same class.
- There are no remaining attributes for further partitioning—majority class is the leaf.
- There are no examples left.

2.4.7.6   Advantages

- It is robust to errors and if the training data contains error—DT algorithms will be best suited to address such problems.
- It is very instinctive and can be explained to anyone with ease.
- Data type is not a constraint as they can handle both categorical and numerical variables.
- It does not require any assumption about the linearity in the data and hence can be used where the parameters are related nonlinearly.
- There are no assumptions about the structure and space distribution.
- It saves data preparation time, as they are not sensitive to missing values and outliers.

2.4.7.7   Disadvantages

- More number of decisions in a tree, less is the accuracy of any expected outcome.
- DTs do not fit well for continuous variables and result in instability and classification plateau.
- DTs are easy to use when compared to other decision-making models but creating large DTs that contain several branches is a complex and time-consuming task.
- DT machine learning algorithms consider only one attribute at a time and might not be best suited for actual data in the decision space.
- Unconstrained, individual trees are prone to over-fitting, but this can be alleviated by ensemble methods (discussed next).

Single DTs are used very rarely, but in comparison with many others, they build very efficient algorithms such as Random Forest or Gradient Tree Boost. There are a couple of algorithms to build DTs, e.g., CART (Classification and Regression Trees, ID3 (Iterative Dichotomiser 3), etc.

Sample Python Code for Decision Tree

Documentation of sklearn DecisionTreeClassifier:

https://scikit-learn.org/stable/modules/generated/sklearn.tree.
DecisionTreeClassifier.html

```
from sklearn.tree import DecisionTreeClassifier
from sklearn.metrics import accuracy_score

# read the train and test dataset
train_data = pd.read_csv('train-data.csv')
test_data = pd.read_csv('test-data.csv')

Create the object of the Decision Tree model

model = DecisionTreeClassifier()

# fit the model with the training data
model.fit(train_x,train_y)

# predict the target on the test dataset
predict_test = model.predict(test_x)
print('Target on test data',predict_test)

# Accuracy Score on test dataset
accuracy_test = accuracy_score(test_y,predict_test)
print('accuracy_score on test dataset : ', accuracy_test)
```

2.4.8   Ensemble Learning

An ensemble method is a technique that combines the predictions of several base estimators built with a given learning algorithm in order to improve generalizability/ robustness over a single estimator [17]. A model that comprises many models is called an ensemble model. The general framework is shown in Fig. 2.9.

2.4.8.1   Types of Ensemble Learning

- Boosting
- Bootstrap aggregation (bagging)

We will discuss each method now.

Boosting

Boosting refers to a group of algorithms that utilize weighted averages to fine tune weak learners into stronger learners. It is all about "teamwork." Each model that runs, dictates what features the next model will focus on. In boosting, one is learning from other which in turn boost the learning.

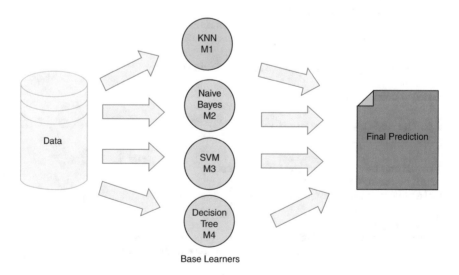

Fig. 2.9 Ensemble learning method

Bootstrap Aggregation (Bagging)

Bootstrap refers to random sampling with replacement. Bootstrap allows us to better understand the bias and the variance with the dataset. Bootstrap involves random sampling of small subset of data from the dataset.

It is a general procedure to reduce the variance for those algorithms that have high variance. Bagging makes each model run independently and then aggregates the outputs at the end without preference to any model.

2.4.9   Random Forests

DTs are sensitive to the specific data on which they are trained. If the training data is changed, the resulting DT will be quite different and in turn the *predictions will be quite different*. Also, DTs are computationally *expensive to train*, carry a big risk of *over-fitting*, and tend to find local optima because they cannot trace back, i.e., there is no back tracking after they have split and advanced. To address these weaknesses, we turn to Random Forest, which illustrates the power of combining many DTs into one model.

Random Forest is one of the most versatile and popular machine learning algorithms. It can be used for both classification and regression problems in ML. It is based on the concept of ensemble learning, which is a process of combining multiple classifiers to solve a complex problem and to improve the performance of the model. As the name suggests, Random Forest is a classifier that contains a number of DTs on various subsets of the given dataset and takes the average to improve the predictive accuracy of that dataset. Instead of relying on one DT, the Random Forest takes the prediction from each tree and based on the majority votes of predictions, and it predicts the final output. The greater number of trees in the forest leads to higher accuracy and prevents the problem of over-fitting.

With its built-in ensemble capacity, the task of building a decent generalized model (on a dataset) is much easier. Random Forests or "random decision forests" is an ensemble learning method called bootstrap aggregation or bagging, combining multiple algorithms to generate better results for classification, regression, and other tasks. Each individual model is weak, but when combined with others, can produce excellent results. As the name of the algorithm, this ML algorithm creates a forest and makes it somehow random.

Random Forest is a tree-based algorithm that involves building several trees (DTs), then combining their output to improve generalization ability of the model. The method of combining trees is known as an ensemble method. The method entails combining weak learners (individual trees) to produce a strong learner. Each tree attempts to create rules in such a way that the resultant terminal nodes could be as pure as possible. Higher the purity, lesser is the uncertainty in arriving at the decision. Random Forest is mostly used in banking (identification of loan risk), medicine (disease trends and risk of identification of disease), marketing (marketing trends), etc.

Random Forest is a popular way to use tree algorithms to achieve good accuracy as well as overcoming the over-fitting problem encountered in single DT algorithm. It also helps to identify most significant features. Random Forest is highly scalable to any number of dimensions and has generally given quite acceptable performance. With Random Forest however, learning may be slow (depending on the parameterization), and it is not possible to iteratively improve the generated models.

The algorithm starts with a "DT" (a tree-like graph or model of decisions) and with node at the top receiving input. It then traverses down the tree, with data being segmented into smaller and smaller sets, based on specific variables. The forest that it builds is an ensemble of DT, and most of the time, it is trained with the "bagging" method. The basic concept behind bagging method is that a combination of learning models increases the overall result. Bagging can be used to reduce the variance for those algorithms that have high variance, typically DTs. Bagging makes each model run independently and then aggregates the outputs at the end without preference to any specific model. There is no interaction between trees during the build ups. A Random Forest is, therefore, considered to be a meta-estimator (i.e., it combines the results of many prediction), which aggregates many DTs.

For classifying a new observation, each tree gives a classification, and the forest chooses the classification having the most votes (over all the trees in the forest). For regression, it is the average of all the trees output.

Each tree is planted and grown as follows: if the number of cases in the training set is N, then the sample of N cases is taken at random but with replacement. This sample will be the training set for growing the tree. Whereas if there are M input features/variables, then a number $m < M$ is specified such that at each node, m variables are selected at random out of the M and the best split on this m is used to split the node. The value of m is held constant during the forest growing. Each tree is grown to the maximum extent possible. Moreover, each tree is grown on a different sample of different data. Since Random Forest has the features to calculate out of bag (OOB) error internally, a cross validation does not make much sense in Random Forest. The number of features that can be used to split on at each node is limited to some percentage of the total (which is known as the hyper parameter). This ensures that the ensemble model does not rely too heavily on some individual feature and makes fair use of all potentially predictive features. Each tree draws a random sample from the original dataset when generating its splits, adding a further element of randomness that prevents over-fitting. By reducing the features to a random subset that may be considered at each split point, it forces each DT in the ensemble to be more different. The effect is that the predictions and, in turn, prediction errors made by each tree in the ensemble are relatively less correlated. When the predictions from these less correlated trees are averaged to make a prediction, it often results in better performance than the bagged DTs. Various hyper parameters are number of samples, number of features, number of trees, and tree depth. Figure 2.10 explains the working of the Random Forest algorithm.

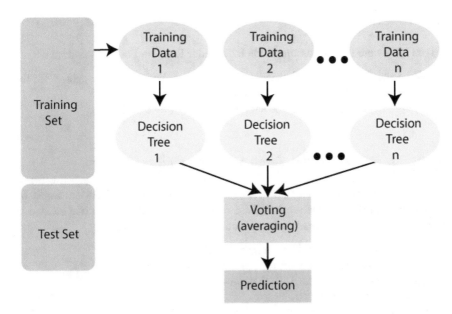

Fig. 2.10 Random Forest algorithm

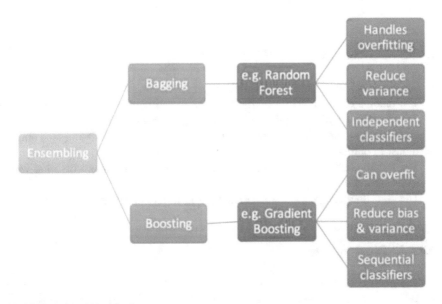

Fig. 2.11 Ensemble methods

Random Forest is a bagging technique and not a boosting technique. The trees in random forests are run in parallel. There is no interaction between these trees while building the trees. Figure 2.11 shows the two types.

2.4.9.1 Assumptions

Given below are the two assumptions for a better Random Forest classifier:

- There should be some actual values in the feature variable of the dataset for the classifier to predict more accurate results.
- The predictions from each tree should have very low correlations.

2.4.9.2 How Does Random Forest Algorithm Work?

Random Forest works in two phases. The first phase creates the random forest by combining N DTs, and the second phase makes predictions for each tree created in the first phase. The process can be explained in the following steps:

1. Select random K data points from the training set.
2. Build the DTs associated with the selected data points (subsets).
3. Choose the number N for DTs that we want to build.
4. Repeat steps 1 and 2.
5. For new data points, find the predictions of each DT, and assign the new data points to the category that wins the majority votes and average of all the output for regression.

2.4.9.3 Advantages

- The method has not encountered over-fitting problem.
- The method can be used for both classification and regression tasks.
- Relatively the method requires very little preprocessing of data.
- Generally, the method maintains accuracy in the presence of missing data and is robust to outliers.
- The method has an implicit feature selection as it gives estimates on what variables are important.
- The algorithm can be parallelized resulting in faster convergence.

2.4.9.4 Disadvantages

- The method is easy to use but analyzing it is difficult.
- When the method generates large number of trees, then it can slowdown making real-time predictions.
- The algorithm gets biased in favor of those features that have more categories or levels. In such situations, variable importance scores do not seem to be reliable.

Random Forest for regression also works in a similar way. Only difference is that output of various trees is averaged to get the predicted output.

Sample Python Code for Random Forest

```
Documentation of sklearn RandomForestClassifier:

https://scikit-learn.org/stable/modules/generated/sklearn.
ensemble.RandomForestClassifier.html

from sklearn.ensemble import RandomForestClassifier
from sklearn.metrics import accuracy_score

Create the object of the Random Forest model

rf_model = RandomForestClassifier()

# fit the model with the training data
rf_model.fit(train_x,train_y)

# number of trees used
print('Number of Trees used : ', rf_model.n_estimators)

# predict the target on the test dataset
predict_test = model.predict(test_x)
print('\nTarget on test data',predict_test)

# Evaluating the Algorithm
accuracy_test = accuracy_score(test_y,predict_test)
print('\naccuracy_score on test dataset : ', accuracy_test)
print(confusion_matrix(test_y,predict_test))

# Variable Importance
feature_importances = pd.DataFrame(rf.feature_importances_,
                  index = train_x.columns,
                  columns=['importance']).sort_values('importance',
ascending=False)
feature_importances
```

2.4.10   K-Means Clustering Algorithm

Sometimes the goal is to assign labels according to the features with no initial information about labels. This task is achieved by creating clusters and assigning labels to each cluster. For example, clustering can be used when there is a large group of users, and there is a need to divide them into groups based on identifiable attributes. In other words, we try to find homogeneous subgroups within the data such that data points in each cluster are as similar as possible. Euclidean-based distance or correlation-based distance is usually used as a similarity measure. The decision of which similarity measure to use is application-specific.

K-means is probably one of the better known and frequently used algorithms. It uses an iterative refinement method to produce its final clustering based on the number of clusters defined by the user (represented by the variable K) and the dataset. For example, if you set K equal to 3, then the dataset will be grouped in three clusters.

K-means is a general-purpose algorithm that makes clusters based on geometric distances between points. It is a centroid-based algorithm that means points are grouped in a cluster according to the distance (mostly Euclidean) from centroid. It is the most widely used centroid-based clustering algorithm. Centroid-based algorithms are efficient but sensitive to initial conditions and outliers. K-means is an efficient, effective, and simple clustering algorithm.

K-means is a non-deterministic iterative method. K-means starts off with arbitrarily chosen data points as proposed means of the data groups and iteratively recalculates new means in order to converge to a final clustering of the data points. The clusters are grouped around centroids, causing them to be globular and have similar sizes. Typically mean is taken to define a centroid representing the center of the cluster. The centroid might not necessarily be a member of the dataset.

The initial result of running this algorithm may not be the best possible outcome and rerunning it with different randomized starting centroids might provide a better performance (different initial objects may produce different clustering results). For this reason, it is a common practice to run the algorithm multiple times with different starting points and evaluate different initiation methods. But another question arises: how does one know the correct value of K, or how many centroids to create? There is no universal answer. Although the optimal number of centroids or clusters is not known a priori, different approaches exist to try to estimate it. One commonly used approach is testing different numbers of clusters and measure the resulting sum of squared errors, choosing the K value at which an increase will cause a very small decrease in the error sum, while a decrease will sharply increase the error sum. The point that defines optimal number of clusters is known as the elbow point and can be used as a visual measure to find the best pick for the value of K.

Because clustering is unsupervised (i.e., there is no "right answer"), data visualization is usually used to evaluate results. If there exists a "right answer" (i.e., we have pre-labeled groups in the training data), then classification algorithms are typically more appropriate.

K-means can be applied to data that has a smaller number of dimensions, is numeric, and is continuous. For example, it has been successfully used for document clustering, identifying crime-prone areas, customer segmentation, insurance fraud detection, public transport data analysis, clustering of IT alerts, etc.

2.4.10.1   Assumptions

- *K*-means assumes that the variance of the distribution of each attribute (variable) is spherical.
- All the constituent variables have the similar variance.
- The prior probability for all *K* clusters is the same, i.e., each cluster has roughly equal number of observations.

If any of these three assumptions is violated, then *K*-means will fail.

2.4.10.2   How Does *K*-Means Algorithm Work?

Given *K* (the number of clusters), the *K*-means algorithm works as follows:

1. Randomly choose *K* data points (seeds) to be the initial centroids, cluster centers.
2. Compute the sum of the squared distance between data points and all centroids.
3. Assign each data point to the closest centroid.
4. Re-compute the centroids for the clusters by taking the average of all data points that belong to each cluster.
5. If a convergence criterion is not met, go to 3.

2.4.10.3   Convergence Criterion

Most of the convergence happens in the first few iterations. Essentially, there are essentially three stopping criteria that can be adopted to stop the *K*-means algorithm:

- Centroids of newly formed clusters do not change.
- Points remain in the same cluster.
- Maximum number of iterations are reached.

We can stop the algorithm if the centroids of newly formed clusters are not changing. Even after multiple iterations, if we are getting the same centroids for all the clusters, we can say that the algorithm is not learning any new pattern, and it is a sign to stop the training.

Another clear sign is that we should stop the training process if the points remain in the same cluster even after training the algorithm for multiple iterations.

Finally, we can stop the training if the maximum number of iterations is reached. Suppose if we have set the number of iterations as 100. The process will repeat for 100 iterations before stopping.

Figure 2.12 shows the basic flow of the algorithm.

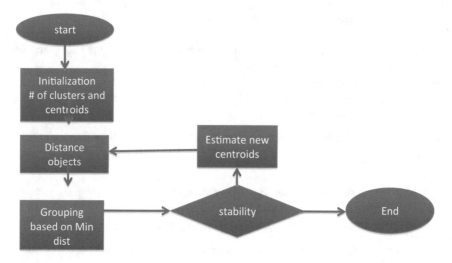

Fig. 2.12 Basic flow of *K*-means algorithm

2.4.10.4 Advantages

- It is fast, simple, and flexible.
- Given a smaller value of *K*, *K*-means clustering computes faster than hierarchical clustering for large number of variables.

2.4.10.5 Disadvantages

- The method requires specification of number of clusters in advance that may not be easy to do.
- The method produces poor clusters if data does not have globular clusters.

Sample Python Code for *K*-Means

Documentation of sklearn *K*-Means:

https://scikit-learn.org/stable/modules/generated/sklearn.cluster.KMeans.html

from sklearn.cluster import KMeans

Create the object of the *K*-Means model with three clusters

model = KMeans(n_clusters=3)

```
# fit the model with the training data
model.fit(train_data)

# Number of Clusters
print('\nDefault number of Clusters : ',model.n_clusters)

# predict the target on the test dataset
predict_test = model.predict(test_data)
```

There are other clustering algorithms also (Hierarchical, DBSCAN etc.) that the reader is referred to [5].

2.4.11 Artificial Neural Networks

Artificial neural networks (ANNs) are inspired by biological systems, such as the brain, and how they process information. Neural networks flourished in the mid-1980s due to their inherent parallel and distributed processing capability. But research in this field was impeded by the ineffectiveness of the back-propagation training algorithm that is widely used to optimize the parameters of neural networks. In recent years, new and improved training techniques such as unsupervised pre-training and layer-wise greedy training have led to a resurgence of interest in neural networks. Increasingly powerful computational capabilities, such as graphical processing unit (GPU) and massively parallel processing (MPP), have also spurred the revival of neural networks. The resurgent research in neural networks has given rise to the invention of models with thousands of layers, i.e., shallow neural networks have evolved into deep learning neural networks. Deep neural networks have been very successful for supervised learning. When used for speech and image recognition, deep learning algorithms perform sometimes better than even humans. Applied to unsupervised learning tasks, such as feature extraction, deep learning also extracts features from raw images or speech with much less human intervention.

ANNs have interconnected processing elements that work in unison to solve specific problems. Neural network consists of three parts: input layer, hidden layer, and output layer as shown in Fig. 2.13. The training samples define the input and output layers. Hidden layers between inputs and outputs are used in order to model intermediary representation of the data that other algorithms cannot easily learn. The number of hidden layers defines the model complexity and model capacity.

ANN comprises "units" arranged in a series of layers, each of which connects to layers on either side. With ANN, extremely complex models can be modeled and can be utilized as a kind of black box, without playing out unpredictable complex feature engineering before training the model. Joined with the "deep approach," even more unpredictable models can be picked up to realize new possibilities. For example, object recognition based on geometry has enhanced enormously utilizing deep neural networks. Applied to unsupervised learning tasks, such as feature extraction.

Fig. 2.13 A typical neural networks architecture

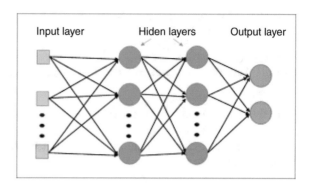

Deep learning has also been used to extract from raw images or speech with much less human intervention. ANNs also learn by example and through experience, and they are extremely useful for modeling nonlinear relationships in high-dimensional data or where the relationship among the input variables is difficult to understand. ANNs can be used for both regression and classification.

2.4.11.1 Advantages

- It performs very well on image, audio, and text data.
- The architecture can be adapted to many complex problems.
- The hidden layers reduce the need for feature engineering.

2.4.11.2 Disadvantages

- It requires very large amount of data.
- It is computationally intensive to train.
- It requires much more expertise to tune (set of architecture and hyper-parameters).

For a more detailed discussion of these algorithms, the readers may see refer to [4–6].

2.5 Usage of ML Algorithms

ML is a huge field, and the ML algorithms described above are just a few examples. The application and choice of an algorithm depends on the kind of problem being attempted. It is hard to know right at the beginning which algorithm will work best. It is usually better to work iteratively. Among the ML algorithms, one identifies as potentially superior/better approaches.

Try each with data, and at the end evaluate the performance of the algorithms to select the best one(s). Finally, developing the right solution to a real-life problem requires awareness of business demands, rules and regulations, budget and stake-holder's concerns, as well as considerable expertise. The executive's guide to AI by McKinsey [9] is a very good reference. Table 2.1 summarizes various algorithms and their usage. Figure 2.14 shows the usage of various algorithms.

2.6 Performance Metrics of ML Algorithms

Usually following the feature engineering step, we select and implement a model to get output. The next step is to find out how effective is the model based on some metric using a test dataset. Evaluating machine learning model is an essential part of every data science project. There are various metrics that can be used to evaluate the performance of ML algorithms such as classification and regression algorithms. Model evaluation metrics are used to assess goodness of fit between model and data. These metrics may also be used to compare different models and select a model. Such evaluation helps to predict how good are these models? The metrics that is chosen to evaluate machine learning model is very important. Choice of metrics influences how the performance of machine learning algorithms is measured and compared. So, we should carefully choose the metrics for evaluating ML perfor-mance for the following reasons:

- How the performance of ML algorithms is measured and compared will be dependent entirely on the metric chosen.
- How the importance of various characteristics weighs in the result will be influenced completely by the choice of metric.

2.6.1 Testing Data

The next important question while evaluating the performance of a machine learning model is what dataset should be used to evaluate model performance. The machine learning model cannot be simply tested using the training set, because the output will be prejudiced, as the process of training the machine learning model has already tuned the predicted outcome to the training dataset. To estimate the generalization error, the model is required to test a dataset that has not been seen yet; identified as a testing dataset. Therefore, for the purpose of testing the model, we would require a labeled dataset. This can be achieved by splitting the training dataset into training dataset and testing dataset. This can be achieved by various techniques such as *k*-fold cross validation, jackknife resampling, and bootstrapping. Techniques such as A/B testing are used to measure performance of machine learning models in production against response from real user interaction.

Table 2.1 ML use cases

Algorithms	Sample business use cases/applications
Linear regression	• Predict monthly gift card sales and improve yearly revenue projections • Predicting the temperature • Understand product-sales drivers such as competition prices, distribution, and advertisement • Optimize price points and estimate product-price elasticity
Logistic regression	• Classify customers based on how likely they are to repay a loan • Predict if a skin lesion is benign or malignant based on its characteristics (size, shape, color, etc.) • Classifying words as nouns, pronouns, and verbs • In voting applications to find out whether voters will vote for a particular candidate or not
Decision Tree	• Provide a decision framework for hiring new employees • Understanding product attributes that make a product most likely to be purchased • Predicting and reducing customer churn across many industries • Fraud detection in the insurance sector • Option pricing in finance
Naïve Bayes	• Analyze sentiment to assess product perception in the market • Create classifiers to filter spam emails • Sentiment analysis and text classification • Recommendation systems like Netflix, Amazon
Support vector machine	• Predict how many patients a hospital is likely to serve in a given time period • Predict how likely someone is to click on an online ad • Comparing the relative performance of stocks over a period of time
Random Forest	• Predict call volume in call centers for staffing decisions • Predict power usage in an electrical distribution grid • Predict part failures in manufacturing • Predict patients for high risks • Predict the average number of social media shares and performance scores • Image and text classification
K-means clustering	• Segment customers into groups by distinct characteristics (e.g., age group)—for instance, to better assign marketing campaigns or prevent churn • Handwriting detection applications and image/video recognition tasks
Hierarchical clustering	• Cluster loyalty-card customers into progressively more micro-segmented groups • Inform product usage/development by grouping customers mentioning keywords in social media data
Reinforcement learning	• Optimize the trading strategy for an option-trading portfolio • Balance the load of electricity grids in varying demand cycles • Stock and pick inventory using robots • Optimize the driving behavior of self-driving cars • Optimize pricing in real time for an online auction of a product with limited supply

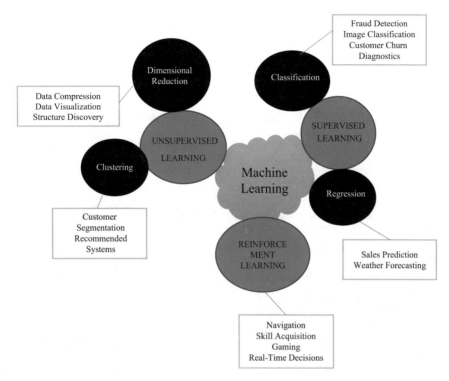

Fig. 2.14 ML usage

2.6.2   Performance Metrics for Classification Models

We have discussed various classification algorithms earlier in the chapter. Here, we discuss various performance metrics [18–20], which can be used to evaluate performance of classification problems. Most of the times classification accuracy is used to measure the performance of a classifier. We begin our discussion with confusion matrix.

2.6.2.1   Confusion Matrix

The confusion matrix is used to have more complete picture when assessing the performance of a model. Confusion matrix gives us a matrix output and describes the complete performance of the model. It is the easiest way to measure the performance of a classification problem where the output can be of two or more type of classes. A confusion matrix is an $N \times N$ matrix, where N is the number of classes being predicted. For the problem in hand where $N = 2$, a confusion matrix is a table with two dimensions viz. "Actual" and "Predicted" and furthermore, both the dimensions have "True Positives (*TP*)," "True Negatives (*TN*)," "False Positives (*FP*)," "False Negatives (*FN*)" as shown in Fig. 2.15.

Fig. 2.15 Confusion matrix

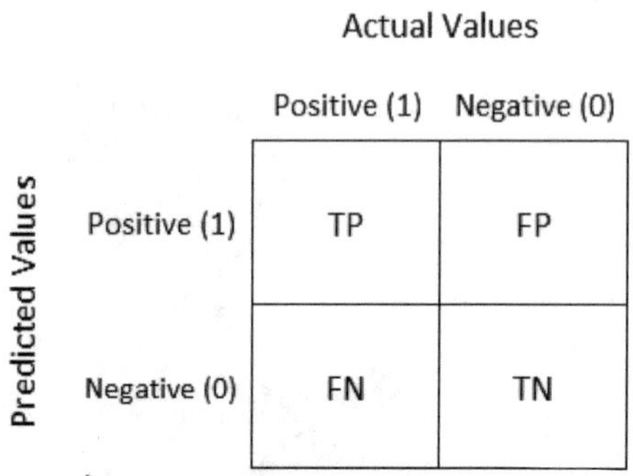

True Positives (TP): It is the case when both actual class and predicted class of data point is 1.

True Negatives (TN): It is the case when both actual class and predicted class of data point is 0.

False Positives (FP): It is the case when actual class is 0 and predicted class of data point is 1. It is also known as type I error.

False Negatives (FN): It is the case when both actual classes is 1 and predicted class of data point is 0. It is also known as type II error.

Confusion matrix shown in Fig. 2.15 forms the basis for measuring accuracy, recall, precision, specificity, etc.

Accuracy: It is the most common performance metric. It is defined as the number of correct predictions made as a ratio of all the predictions made. It can be calculated from the confusion matrix as follows:

$$Accuracy = \frac{TP + TN}{TP + FP + FN + TN}$$

It works well only if there are almost equal number of samples belonging to each class. The real problem arises, when the cost of misclassification of the minor class samples are very high. If we deal with a rare but fatal disease, the cost of failing to diagnose the disease of a sick person is much higher than the cost of sending a healthy person to more tests.

Classification Report: This report consists of the scores of Precisions, Recall, F1, and Support. They are explained as follows:

Precision: It is the number of correct positive results divided by the number of positive results predicted by the classifier.

$$Precision = \frac{TP}{TP + FP}$$

Recall or Sensitivity: It is the number of correct positive results divided by the number of **all** relevant samples (all samples that should have been identified as positive).

Table 2.2 Various performance metrics

Metric	Formula	Interpretation
Accuracy	$\frac{TP+TN}{TP+TN+FP+FN}$	Overall performance of the model
Precision	$\frac{TP}{TP+FP}$	How accurate positive predictions are
Recall (sensitivity)	$\frac{TP}{TP+FN}$	Coverage of actual positive sample
Specificity	$\frac{TN}{TN+FP}$	Coverage of actual negative sample
F1 score	$\frac{2TP}{2TP+FP+FN}$	Hybrid metric useful for unbalanced classes

$$\text{Recall} = \frac{TP}{TP + FN}$$

Specificity: Specificity is defined as the number of negatives returned by our ML model.

$$\text{Specificity} = \frac{TN}{TN + FP}$$

F1 Score: This score will give us the harmonic mean of precision and recall. Mathematically, the range for *F1* score is [0.0.1]. It tells how precise classifier is (how many instances it classifies correctly), as well as how robust it is (it does not miss significant number of instances). High precision but lower recall gives an extremely accurate prediction, but then it misses a large number of instances that are difficult to classify. Greater the *F1* score, better is the performance of the model. *F1* score tries to find a balance between precision and recall. We can calculate *F1* score with the help of following formula:

$$F1 = 2 * \frac{\text{Precision} * \text{Recall}}{(\text{Precision} + \text{Recall})}$$

It is difficult to compare two models with low precision and high recall or vice versa. So, to make them comparable, we use *F*-Score. *F*-score helps to measure Recall and Precision at the same time. It uses Harmonic Mean in place of Arithmetic Mean by punishing the extreme values more. This is so because it weighs equally the relative contribution of precision and recall.

This seems simple. However, there are situations for which one would like to give more importance/weight to either precision or recall. Altering the above expression for F1 score requires including an adjustable parameter beta. The expression then becomes:

$$F_\beta = \left(1 + \beta^2\right) \frac{\text{Precision} \cdot \text{Recall}}{\left(\beta^2 \cdot \text{Precision}\right) + \text{Recall}}$$

All the above metrics are summarized in Table 2.2.

Table 2.3 TPR and FPR

Metric	Formula	Equivalent
True-positive rate (TPR)	$\frac{TP}{TP+FN}$	Recall, sensitivity
False-positive rate (FPR)	$\frac{FP}{TN+FP}$	$1-$specificity

Fig. 2.16 Receiver
operating characteristic
(ROC)

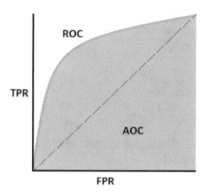

Receiver Operating Curve (ROC): ROC is the plot of true-positive rate versus false-positive rate by varying the threshold. These metrics are summed up in Table 2.3.

Area Under ROC Curve (AUC): The area under the ROC also known as AUC is the area under the ROC as shown in Fig. 2.16. It is a performance metric, based on varying threshold values. AUC of a classifier is equal to the probability that the classifier will rank a randomly chosen positive example higher than a randomly chosen negative example.

In other words, ROC is a probability curve, and AUC measures the separability. In simple words, AUC-ROC metric describes the capability of model in distinguishing the classes. Higher the AUC, better is the model.

Mathematically, it can be created by plotting TPR (true-positive rate), i.e., sensitivity or recall vs. FPR (false-positive rate), i.e., 1-specificity, at various threshold values. Following is the graph showing ROC, AUC having TPR at *y*-axis and FPR at *x*-axis.

Gini coefficient: The Gini coefficient is sometimes used in classification problems. Gini $= 2 * $ AUC $- 1$, where AUC is the area under the curve (see the *ROC curve* entry above). A Gini ratio above 60% corresponds to a good model. Not to be confused with the Gini index or Gini impurity, used when building DTs.

LOGLOSS (logarithmic loss): It is also called logistic regression loss or cross-entropy loss. It is basically defined on probability estimates and measures the performance of a classification model where the input is a probability value between 0 and 1. It offers better insight by differentiating it with accuracy. As we know that accuracy is the count of predictions (predicted value = actual value) in our model whereas Log Loss is the amount of uncertainty of our prediction based on how much it varies from the actual label. With the help of Log Loss value, we can have more accurate view of the performance of our model.

Example:

The following is a simple recipe in Python which will give an insight about how we can use the above performance metrics on two class model.

```
from sklearn.metrics import confusion_matrix from sklearn.
metrics import accuracy_score from sklearn.metrics import
classification_report from sklearn.metrics import roc_auc_score
from sklearn.metrics import log_loss
Y = [1, 1, 0, 1, 0, 0, 1, 0, 0, 0]
Y_pred = [1, 0, 1, 1, 1, 0, 1, 1, 0, 0]
results = confusion_matrix(Y, Y_pred)
print ('Confusion Matrix :')
print (results)
print ('Accuracy Score is',accuracy_score(Y, Y_pred)) print
('Classification Report : ')
print (classification_report(Y, Y_pred))
print ('AUC-ROC:',roc_auc_score(Y, Y_pred))
print ('LOGLOSS Value is',log_loss(Y, Y_pred))
```

Output
Interpretation of results

```
Confusion Matrix :
[[3 3]
 [1 3]]
```

```
Accuracy Score is 0.6
Classification Report :
          precision   recall   f1-score   support
0           0.75       0.50      0.60        6
1           0.50       0.75      0.60        4
micro avg      0.60     0.60       0.60      10
macro avg      0.62     0.62       0.60      10
weighted avg      0.65   0.60       0.60    10
```

```
AUC-ROC: 0.625
LOGLOSS Value is 13.815750437193334
```

2.6.2.2   Regression Metrics

Here we are going to discuss various performance metrics that can be used to evaluate regression models.

Table 2.4 Additional performance metrics for regression

Mallow's Cp	Akaike information criterion (AIC)	Bayesian information criterion (BIC)	Adjusted R^2
$\frac{SS_{res}+2(p+1)\sigma^2}{n}$	$2[(p+2) - \log(L)]$	$Log(n)(p+2) - 2\log(L)$	$1 - \left(\frac{1-R^2(n-1)}{n-p-1}\right)$

Mean absolute error (MAE): It is the average of sum of the absolute difference between the predicted and actual values. Note that MAE does not distinguish the direction of the performance error, i.e., there is no indication about underperformance or overperformance of the model. It can be calculated as:

$$\text{MAE} = \frac{1}{n}\sum_{i=1}^{n} |\, y_i - y_{\text{pred}}\,|$$

where y_i is actual value and y_{pred} is the predicted value.

Mean squared error: Mean squared error (MSE) is quite similar to mean absolute error, the only difference being that MSE takes the average of the sum of the squares of the difference between the original values and the predicted values. The advantage of MSE is that it is easier to compute the gradient. Whereas, mean absolute error requires complicated linear programming tools to compute the gradient. As, we take square of the error, the effect of larger errors become more pronounced than smaller error, and hence, the model can now focus more on the larger errors. It can be calculated as:

$$\text{MSE} = \frac{1}{n}\sum_{i=1}^{n} \left(y_i - y_{\text{pred}}\right)^2$$

Coefficient of determination: R^2 metric is generally used for explanatory purpose and provides an indication of the goodness of fit of predicted values to the actual output values, coefficient of determination, often noted as R^2 provides a measure of how well the observed outcomes are replicated by the model and is defined as:

$$R^2 = 1 - \frac{SS_{\text{res}}}{SS_{\text{tot}}}$$

The following metrics are also used to assess the performance of the regression models, by taking into account the number of variables/predictors p that they take into modeling (Table 2.4):

Where L is the likelihood and σ^2 is an estimate of the variance associated with each response.

Example

```
from sklearn.metrics import r2_score
from sklearn.metrics import mean_absolute_error
from sklearn.metrics import mean_squared_error
Y = [5, -1, 2, 10]
Y_pred = [3.5, -0.9, 2, 9.9]
print ('R Squared =',r2_score(Y, Y_pred))
print ('MAE =',mean_absolute_error(Y, Y_pred))
print ('MSE =',mean_squared_error(Y, Y_pred))
```

Output
Some interpretation
R Squared = 0.9656060606060606
MAE = 0.4249999999999993
MSE = 0.5674999999999999

2.7   Most Popular Machine Learning Software Tools

There are several ML software products that are available. Table 2.5 shows the most popular ones among them [21].

Table 2.5 ML software tools

Name	Platform	Cost	Written in language	Features
Scikit Learn	Linux, Mac, Windows	Free	Python, Cython, C, C++	Classification, regression, clustering, preprocessing, etc.
PyTorch	Linux, Mac, Windows	Free	Python, C++, CUDA	Autograd module, nn module
TensorFlow	Linux, Mac, Windows	Free	Python, C++, CUDA	Deep learning, ML
WEKA	Linux, Mac, Windows	Free	JAVA	Classification, regression, clustering, preprocessing, etc.
KNIME	Linux, Mac, Windows	Free	JAVA	Supports text mining and image processing
Apache Mahout	Cross platform	Free	JAVA Scala	
Keras.io	Cross platform	Free	Python	API for neural networks
Rapid Miner	Cross platform		JAVA	Data loading and transformation, etc.

2.8 Machine Learning Platforms

Machine learning platforms are software products that enable data scientists and developers to create, deploy, and manage their own advanced analytics models. Data science platform is defined [9] as:

"A cohesive software application that offers a mixture of basic building blocks essential for creating all kinds of data science solution, and for incorporating those solutions into business processes, surrounding infrastructure and product."

ML platforms are not the wave of the future anymore. They are happening now. Developers need to know how and when to harness their power. A good data science and machine learning platform should offer data scientists all the building blocks for creating a solution. It should also provide the experts with an environment where they can incorporate the solutions into products and business processes. The platform needs to provide data scientists with all the support needed for the tasks such as data access, data preparation, visualization, interactive exploration, deployment, and performance engineering.

Data scientists use a data science and machine learning platform that enables them to work both online and offline. With the introduction of cloud-based platforms, data scientists can now work with their data on any Internet-enabled device. They can also share components of their work with their colleagues or collaborate with them securely on certain tasks. Besides having cloud features, the data science and machine learning platform should also run faster to provide accurate results. Several software vendors are currently unleashing out software products that match this description. In this section, we will discuss some of the well-known and proven platforms available that are currently available.

2.8.1 Alteryx Analytics

Alteryx Analytics [10] provides a machine learning platform for building models in a workflow. Platform can discover, prepare, and analyze all the data, then deploy and share analytics at scale for deeper insights. For a complete list of system requirements and supported data sources, visit [10].

2.8.2 $H_2O.ai$

The company offers H_2O for deep-learning H_2O Steam and H_2O Flow. H_2O.ai was designed for the Python, R, and Java programming languages. Available on Mac, Windows, and Linux, H_2O provides developers with the tools they need to analyze data sets in the Apache Hadoop file systems as well as those in the cloud.

2.8.3   KNIME Analytics Platform

KNIME is an open-source platform useful in enterprises looking to boost their performance, security, and collaboration. A cloud version is available on Microsoft Azure and Amazon AWS. It can blend the data from any source (text formats, databases, and data-ware houses, and from sources such as Twitter, AWS S3, Google sheets, and Microsoft Azure) [11].

2.8.4   RapidMiner

RapidMiner [12] platform comes with RapidMiner Hadoop for extending the platform's execution capabilities to a Hadoop environment, RapidMiner Studio for model development and RapidMiner Server that enables data scientists to share, collaborate on, and maintain models.

2.8.5   Databricks Unified Analytics Platform

Databricks Apache Spark [9]-based Unified Analytics platform offers features for real-time enablement, performance, operations, reliability, and security on AWS. Apache Spark-based platform combines data engineering and data science capabilities that use a variety of open-source languages. Apache Spark MLlib features an algorithms database with a focus on clustering, collaborative filtering, classification, and regression. Developers can find Singa, an open-source framework, that contains a programing tool that can be used across numerous machines to use their deep learning networks. Azure Databricks is an integrated service within Microsoft Azure that provides a high-performance Apache Spark-based platform optimized for Azure.

2.8.6   Microsoft's Azure Machine Learning Studio

Microsoft provides the solution for data science and Ml though its Azure software products. These products include Azure Machine-learning, Power BI, Azure Data Lake, Azure HDInsight, Azure Stream Analytics, and Azure Data Factory. Its cloud-based Azure Machine-learning Studio is ideal for data scientists who want to build test and execute predictive analytics solutions on their data. The cloud platform also offers advantages in terms of performance tuning, scalability and agile support for open-source technology.

2.8.7   Google's Analytics Platform

Google's core ML platform includes Cloud ML Engine, Cloud AutoML, TensorFlow, and BigQuery. Its ML components require other Google components for end-to-end capabilities, such as Google Cloud Dataprep, Google datalab, Google cloud Datproc, and Google Kubernetes Engine. Most of these components require the presence of Google Cloud Platform. Google offers a rich ecosystem of AI products and solutions, ranging from hardware (Tensor Processing Unit [TPU]) and crowdsourcing (Kaggle) to world-class ML components for processing unstructured data like images, videos, and texts. Aided by plethora of online resources, documentation, and tutorials, TensorFlow provides a library that contains data flow graphs in the form of numerical computation. The purpose of this approach is that it allows developers to launch frameworks of deep learning across multiple devices including mobile, tablets, and desktops. Historically, TensorFlow was aimed at "democratizing" machine learning. It was the first platform that made ML simple, visual, and accessible to this degree. The most significant selling pint of TensorFlow is Keras, which is a library for efficiently working with neural networks programmatically.

2.8.8   IBM Watson

IBM's Watson platform is where both business users and developers can find a range of AI tools. Users of the platform can build virtual agents, cognitive search engines, and chatbots with the use of starter kits, sample code, and other tools that can be accessed via open APIs.

2.8.9   Amazon Web Services (AWS)

Amazon Web Services include Amazon Lex, Amazon Rekognition Image, and Amazon Polly. Each one of these is used in a different way by developers to create ML tools. For example, Amazon Polly takes advantage of AI to automate the process of translating voice to written text. Similarly, Amazon Lex forms the basis of the brand's chatbots that are used with its personal assistant, Alexa. There are many other AI services Amazon offers.

The moot point is AI and ML technologies have been commoditized, and there is a race to provide as many features at as low a price as possible. Basically, it all comes down to what stack one has access to and what the goals are.

2.9 Points to Ponder

1. How can one measure the correctness of a ML algorithm?
2. How can one measure the performance of a ML algorithm?
3. What is the meaning of over-fitting and why it is not desirable?

References

1. Chappell, D. (2015). *Introducing azure machine learning: A guide for technical professionals.* San Francisco, CA: Chappell & Associates.
2. https://hackernoon.com/choosing-the-right-machine-learning-algorithm-68126944ce1f.
3. http://blog.echen.me/2011/04/27/choosing-a-machine-learning-classifier/.
4. James, G., Witten, D., Hastie, T., & Tibshirani, R. (2013). *An introduction to statistical learning, with applications in R.* Springer: New York.
5. Mitchell, T. (1997). *Machine learning.* New York, NY: McGraw-Hill Science.
6. Bishop, C. (2006). *Pattern recognition and machine learning.* Springer, New York.
7. https://medium.com/@zxr.nju/what-is-the-kernel-trick-why-it-is-important-98a98db0961d.
8. https://data-flair.training/blogs/svm-kernel-functions/.
9. databricks.com.
10. https://www.alteryx.com/.
11. knime.com/knime-analytics-platform.
12. rapidminer.com.
13. https://www.knowledgehut.com/blog/data-science/machine-learning-algorithms.
14. https://medium.com/@Zelros/a-brief-history-of-machine-learning-models-explainability-f1c3301be9dc.
15. https://medium.com/@aravanshad/how-to-choose-machine-learning-algorithms-9a92a448e0df.
16. https://en.wikipedia.org/wiki/Feature_scaling.
17. http://scikit-learn.org/stable/tutorial/machine_learning_map/index.html.
18. https://developers.google.com/machine-learning/glossary.
19. https://www.analyticsvidhya.com/glossary-of-common-statistics-and-machine-learning-terms/.
20. https://semanti.ca/blog/?glossary-of-machine-learning-terms.
21. https://medium.com/towards-artificial-intelligence/machine-learning-algorithms-for-beginners-with-python-code-examples-ml-19c6afd60daa.

Chapter 3
Deep Learning and Cloud Computing

During recent years, deep learning has acquired the status of a buzzword in the technical community. Literally, every month, we see new developments and trends being reported in the field of AI and ML applications that transform the businesses. Machine learning is imbibed in the majority of business operations and have proved to be an integral part in a decision-making. However, it is artificial intelligence with DL, which amplifies the overall capability in specific domains. The benefits of using models based on these technologies in businesses have brought a significant shift in a way companies are investing and adopting in these technologies. In this chapter, we will be discussing the concept of DL to provide an intuition of how it works.

Currently, AI is advancing at a great pace, and DL is one of the main contributors to that. With unlimited applications like prediction, speech and image processing, and recognition, natural language processing, gene mapping, and more, DL is being extensively used by companies. Incorporating AI and ML will be a paradigm shift for companies who have till date depended on rule-based engines to drive their businesses. This helps in establishing an autonomous and a self-healing process in an organizational growth. Deep leaning has become the main driver of many new applications like self-driving cars, and it is time to really look at why this is the case. It is good to understand the basics of DL as they are changing the world we live in.

Conventional machine learning algorithms/models have always been very powerful in processing structured data and have been widely used by various businesses for customer segmentation, credit scoring, fraud detection, churn prediction, and so on. The success of these models highly depends on the performance of feature engineering phase: the more we work close to the business to extract relevant knowledge from the structured data, the more powerful the model will be.

When it comes to unstructured data (images, text, voice, videos), hand engineered features are time consuming, brittle, and not scalable in practice. That is where neural networks based DL scores. This is due to their ability to automatically discover the representations needed for feature detection or classification from raw data. This replaces manual feature engineering and allows a machine to both learn

P. Gupta, N. K. Sehgal, *Introduction to Machine Learning in the Cloud with Python*, https://doi.org/10.1007/978-3-030-71270-9_3

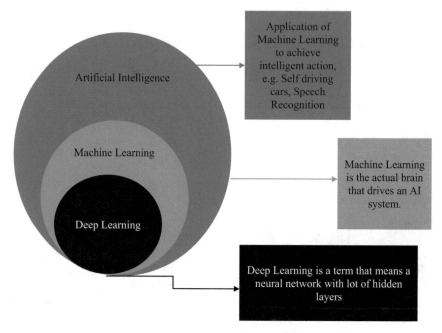

Fig. 3.1 Artificial intelligence and machine learning

the features and use them to perform specific tasks. Improvements in hardware (GPUs) and software (advanced models/research related to AI) also contributed to deepen the learning from data using DL. Figure 3.1 shows the boundary of various AI domains.

3.1 Deep Learning (DL)

DL is a subfield of machine learning in artificial intelligence that deals with algorithms inspired from biological structure and functioning of a brain to aid machines with intelligence. DL is completely based on ANN. In other words, DL is an approach to learning where one can make a machine imitate the network of neurons in a human brain. DL is a way of classifying, clustering, and predicting things by using a network that has been trained using vast amounts of data. Neural network is the main tool in DL. Neural network mimics the human brain. In fact, similar to how we learn from experience, the DL algorithm performs a task repeatedly, each time tweaking it a little bit to improve the performance. Neural networks are set of algorithms, modeled after human brain, that are designed to recognize patterns from historical data. We refer to DL because there are various (deep) layers that enable learning. DL creates many layers of neurons, attempting to learn structured representation of big data, layer by layer. Just about any problem that requires

Table 3.1 Biological
NN vs. ANN

Biological	Artificial
Dendrites	Inputs
Nucleus	Nodes
Synapse	Weights
Axon	Outputs

"thought" to figure out is a problem DL can be used to solve. It enables to assemble complex concepts out of simple concepts. DL tackles the complex problem by breaking the input (mappings) into simpler form which is described by each layer of the model. The models are constructed with connected layers. In between input layer and output layer, there is a set of hidden layers that gave rise to the word deep which means networks that joins neurons in more than two layers.

DL is a particular kind of machine learning that achieves great power and flexibility by learning to represent the world as a nested hierarchy of concepts, with each concept defined in relation to simpler concepts, and more abstract representations computed in terms of less abstract ones. The algorithm has a unique feature, i.e., automatic feature extraction. It means that this algorithm automatically grasps the relevant features required for the solution of the problem. It reduces the burden on the programmer to select the features explicitly. It forms a hierarchy of low-level features. This enables DL algorithms to solve more complex problems consisting of a vast number of nonlinear transformational layers. In DL neural network, each hidden layer is responsible for training the unique set of features based on the output of the previous layer. As the number of hidden layers increases, the complexity and abstraction of data also increases. It can be used to solve supervised, unsupervised or semi-supervised learning.

In DL, we do not need to explicitly program everything. They can automatically learn representations from data such as images, video or text, without introducing hand-coded rules. Their highly flexible architectures can learn directly from raw data and can increase their predictive performance when provided with more data. For example, in face recognition, how pixels in an image create lines and shapes, how those lines and shapes create facial features, and how these facial features are arranged into a face. For example, a DL model known as a convolutional neural network can be trained using a large number (as in millions) of images, such as those containing cats. This type of neural network typically learns from the pixels contained in the images it acquires. For example, it can classify groups of pixels that are representative of a cat's features such as paw, ears, and eyes, indicating the presence of a cat in an image.

The concept of DL is not new. It has been around for several years. It is on hype nowadays because earlier we did not have that much processing power and a lot of data. As in the last 20 years, the processing power increases exponentially, DL and machine learning came into prominence. A formal definition of DL is neurons. Table 3.1 compares biological neural network to artificial neural network.

3.2   Historical Trends

The DL is not a new technology, it dates back to the 1940s. The term DL was introduced to the machine learning community by Rina Dechter in 1986 [1]. We will look into some of the historical facts and trends to understand its origin. There have been three waves of development: DL known as cybernetics (1940s–1960s), DL known as connectionism (1980s–1990s), and the current wave under the name DL (2006 to present) as shown in Fig. 3.2.

The first wave cybernetics started with the development of theories of biological learning and implementation of the first models such as the perceptron, enabling the training of single neuron. The second wave started with the connectionist approach with backpropagation to train a neural network with one or two hidden layers. The current and third wave started around 2006 [2], which we will discuss in the upcoming parts of this series in detail.

3.3   How Do Deep Learning Algorithm Learn?

DL algorithms use neural networks to find associations between a set of inputs and outputs. The basic structure is shown in Fig. 3.3.

A neural network is composed of input, hidden, and output layers—all of which are composed of "nodes" which are fully connected, as shown in Fig. 3.3. This type of structure is known as multilayer perceptron (MLP). MLPs have input and output layers; besides these, they have multiple hidden layers in between the aforementioned layers as shown in Fig. 3.3. Input layers take in a numerical representation of data (e.g., images with pixel specs), output layers output predictions, while hidden

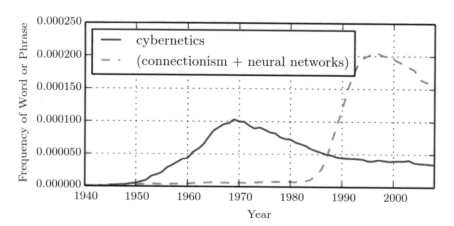

Fig. 3.2 Historical trends of DL. (Sources: Google images)

Fig. 3.3 Multilayer perceptron

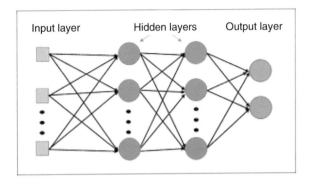

layers are correlated with most of the computation. It is the number of hidden layers that give it the capability of DL.

After the neural network passes its inputs, all the way to its outputs, the network evaluates how good its prediction was (relative to the actual output) through a loss function measured by mean squared error as shown in the equation below:

$$\frac{1}{n}\sum_{i=1}^{n}\left(Y_i - \widehat{Y}_i\right)^2$$

$\widehat{Y}$ represents the predicted value, while Y represents the actual value, and n represents sample count. The goal is to minimize the loss function by adjusting the parameters (weights and biases) of the network. By using backpropagation through gradient descent, the network backtracks through all the layers to update the weights and biases of every node. The continuous updates of the weights and biases of the network ultimately turn it into a more precise function approximator—one that models the relationship between inputs and outputs. The learning algorithm can be described as follows:

1. The inputs are pushed forward through the MLP by taking the dot product of the input with the weights that exist between the input layer and the hidden layer. This dot product yields a value at the hidden layer.
2. MLPs utilize activation functions at each of their calculated layers. There are many activation functions such as sigmoid function, hyperbolic tangent (tanh), and rectified linear unit (ReLU) as shown in Fig. 3.4. Push the calculated output at the current layer using one of these activation functions.
3. Once the calculated output at the hidden layer has been pushed through the activation function, push it to the next layer in the MLP by taking the dot product with the corresponding weights.
4. Repeat steps 2 and 3 until the output layer is reached.
5. At the output layer, the calculations will either be used for a backpropagation algorithm that corresponds to the activation function that was selected for the MLP (in the case of training) or a decision will be made based on the output (in the case of testing).

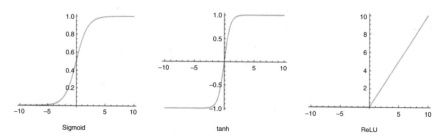

Fig. 3.4 Activation functions

Table 3.2 Activation functions

Activation function	Pros	Cons
Sigmoid	Used in the output layer for binary classification	Output ranges from 0 to 1
Tanh	Better than sigmoid	Updates parameters slowly when points are at extreme ends
ReLu	Updates parameters faster as slope is 1 when $x > 0$	Zero slope when $x < 0$

3.3.1 Activation Functions

The purpose of the activation function is to introduce nonlinearity into the output of a neuron. This is important because most real-world data is nonlinear, and we want neurons to learn these nonlinear representations. Every activation function takes a single number and performs a certain fixed mathematical operation on it. Fig. 3.4 shows various activation functions (Table 3.2):

3.4 Architectures

There are various types of deep neural networks, with structures suited to different types of tasks. For example, Convolutional Neural Networks (CNNs) are typically used for computer vision tasks, while Recurrent Neural Networks (RNNs) [19] are commonly used for processing languages. Each has its own specializations, in CNNs the initial layers are specialized for extracting distinct features from the image, which are then fed into a more conventional neural network to allow the image to be classified. RNNs differ from a traditional feed-forward neural network in that they do not just feed data from one neural layer to the next but also have built-in feedback loops, where data output from one layer is passed back to the layer preceding it—lending the network a form of memory. There is a more specialized form of RNN that includes what is called a memory cell and that is tailored to processing data with lags between inputs.

The most basic type of neural network is a multi-layer perceptron network, the type discussed above in the handwritten figures example, where data is fed forward between layers of neurons. Each neuron transforms the values fed using an activation function, which change these values into a form that, at the end of the training cycle, will allow the network to calculate how far off it is from making an accurate prediction.

There are a large number of different types of deep neural networks. No one network is inherently better than the other. They just are better suited to learning particular types of tasks.

More recently, generative adversarial networks (GANS) are extending use of neural networks. In this architecture two neural networks do battle, the generator network tries to create convincing "fake" data, and the discriminator attempts to tell the difference between fake and real data. With each training cycle, the generator gets better at producing fake data, and the discriminator gains a sharper eye for spotting those fakes. By pitting the two networks against each other during training, both can achieve better performance. GANs have been used to carry out some remarkable tasks, such as turning dashcam videos from day to night or from winter to summer and have applications ranging from turning low-resolution photos into high-resolution alternatives and generating images from written text. GANs have their own limitations, however, that can make them challenging to work with, although these are being tackled by developing more robust GAN variants.

3.4.1 Deep Neural Network (DNN)

For the past decade, deep neural networks have been used in image recognition, speech, and even play games with high accuracy. The name deep neural evolved from the use of many layers making it a "deep" network to learn more complex problems. The success stories of DL have only surfaced in the last few years because the process of training a network is computationally heavy and needs large amount of data. The success of DL found applications when faster computation and massive data storage became more available and affordable.

A DNN is an ANN with multiple hidden layers between the input and output layers. DNN is a neural network with a certain level of complexity (having multiple hidden layers in between input and output layers). They are capable of modeling and processing nonlinear relationships through activation function described earlier in the chapter. The DNN finds the correct mathematical manipulation to turn the input into the output, whether it is a linear relationship or a nonlinear relationship. The network moves through the layers calculating the probability of each output. The user can review the results and select which probabilities the network should display (above a certain threshold, etc.) and return the proposed label. Each mathematical manipulation as such is considered a layer, and complex DNN have many layers,

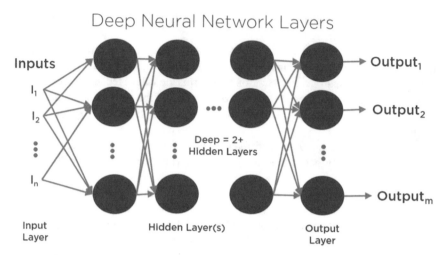

Fig. 3.5 Deep neural network architecture

hence the name "deep" networks. The system must process layers of data between the input and output to solve a task. Deep neural network represents the type of machine learning when the system uses many layers to derive high-level functions from input information. It means transforming the data into a more creative and abstract component.

DNNs are typically feedforward networks in which data flows from the input layer to the output layer without looping back. A simplified version of DNN is represented as a hierarchical (layered) organization of neurons (similar to the neurons in the brain) with connections to other neurons. These neurons pass a signal to other neurons based on the received input and form a complex network that learns with some feedback mechanism. The input data is given to the neurons in the first layer also known as input layer which then provide its output to the neurons within the next layer and so on till the final layer or the output layer. Each layer can have one or many neurons, and each of them will compute a small function, i.e., activation function. If the incoming neurons result in a value greater than some threshold, the output is passed else ignored. The connection between two neurons of successive layers have an associated weight. The weight defined the strength of the input to the output for the next neuron and eventually for the overall final output. Initially weights would be all random but during training, these weights are updated iteratively to learn to predict a correct output. Iterating the process several times step-by-step, with more and more data, helps the network update the weights appropriately to create a system where it can take a decision for predicting the output based on the rules. The prediction accuracy of a network depends on its weights and biases. A typical DNN can be defined by building blocks like a neuron, layers (input, hidden, and output), weights, an activation function, and a learning mechanism (backpropagation). Figure 3.5 represents a typical deep neural network.

3.4.2 *Recurrent Neural Network (RNN)*

A recurrent neural network is a class of artificial neural networks where connections between nodes form a directed graph along a temporal sequence. This allows it to exhibit temporal dynamic behavior. Derived from feedforward neural networks, RNNs can use their internal state (memory) to process variable length sequences of inputs [3]. This makes them applicable to tasks such as unsegmented, connected handwriting recognition [4], or speech recognition [5].

RNNs are neural networks in which data can flow in any direction. The basic concept is to utilize sequential information. In a normal neural network, it is assumed that all inputs and outputs are independent of each other. If we want to predict the next word in a sentence, we have to know which words come before it. RNNs are called recurrent as they repeat the same task for every element of a sequence, with the output being based on the previous computations. RNNs thus can be said to have a memory that captures information about what has been previously calculated. In theory, RNNs can use information in a very long sequences, but in reality, they can look back only a few steps. A typical RNN is shown in Fig. 3.6.

Recurrent neural network remembers the past, and its decisions are influenced by what it has learnt from the past. Note: Basic feedforward networks "remember" things too, but they remember things they learnt during training. For example, an image classifier learns what a "1" looks like during training and then uses that knowledge to classify things in production.

RNNs learn similarly while training, in addition, they remember things learnt from prior input(s) while generating output(s). RNNs can take one or more input vectors and produce one or more output vectors, and the output(s) are influenced not just by weights applied on inputs like a regular NN, but also by a "hidden" state vector representing the context based on prior input(s)/output(s). So, the same input could produce a different output depending on previous inputs in the series. Long

Fig. 3.6 Recurrent neural network

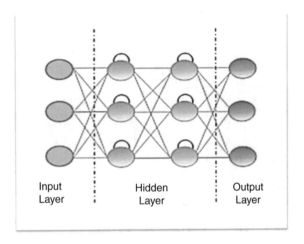

Input Layer   Hidden Layer   Output Layer

short-term memory networks (LSTMs) are most commonly used RNNs. The readers
are referred to [5–7] for more details.

3.4.3   Convolutional Neural Networks (CNN)

Convolutional neural networks also known as CNNS or Convnets are a class of deep
neural networks, most commonly applied to analyzing visual imagery. These networks have been some of the most influential innovation in the field of computer
vision. The year 2012 was the first year that neural nets grew to prominence as Alex
Krizhevsky used them to win that year's ImageNet competition dropping the
classification error record from 26% to 15%, an astounding improvement at the
time [2]. Ever since, a host of companies have been using at the core of their services.
They have applications in image and video recognition, recommender systems,
medical image analysis, natural language processing, and time series. It is mostly
applied to images because there is no need to check all the pixels one by one. CNN
checks an image by blocks, starting from the left upper corner and moving further
pixel by pixel up to a successful completion. Then the result of every verification is
passed through a convolution layer, where data elements have connections while
others do not. Based on this data, the system can produce the result of the verification
and can conclude what is in the picture.

The name "Convolutional Neural Network" indicates that the network employs a
mathematical operation called convolution. Convolution is a specialized kind of
linear operation. Convolutional networks are simply neural networks that use convolution in place of general matrix multiplication in at least one of their layers
[7]. CNNs use relatively little pre-processing compared to other image classification
algorithms. This means that the network learns from the filters that in traditional
algorithms were hand-engineered. This independence from prior knowledge and
human effort in feature design is a major advantage. CNNs are composed of five
basic blocks as shown in Fig. 3.7:

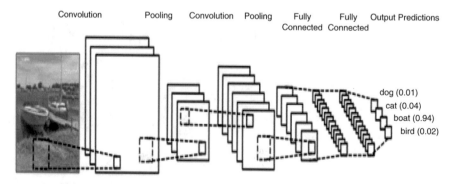

Fig. 3.7 Convolutional neural network architecture

- An input layer
- Convolution layer
- ReLU layer
- Pooling layer
- Fully connected layer

The readers are referred to [7] for more details.

A CNN is a DL algorithm that can take in an input image, assign importance (learnable weights and biases) to various aspects/objects in the image, and be able to differentiate one from the other. The pre-processing required in a CNN is much lower as compared to other classification algorithms. While in primitive methods, filters are hand-engineered, with enough training, CNNs have the ability to learn these filters/characteristics. The architecture of a CNN is analogous to that of the connectivity pattern of neurons in the human brain and was inspired by the organization of the visual cortex. Individual neurons respond to stimuli only in a restricted region of the visual field known as the receptive field. A collection of such fields overlap to cover the entire visual area.

3.5   Choosing a Network

How to choose an appropriate network? One has to decide depending upon the problem being solved, e.g., either building a classifier or trying to find patterns in the data. Following points should be considered while selecting a network [20]:

- For text processing, sentiment analysis, any language model that operates at character level, parsing, and name entity recognition, we use recurrent network.
- For image recognition, use convolution network.
- For speech recognition, we use recurrent network.

In general, multilayer perceptrons or DNN with rectified linear activation function or ReLU are good choice for classification. For time series analysis, it is always recommended to use recurrent network.

3.6   Deep Learning Development Flow

Steps involved in developing a DL solution:

1. Selection of a framework for development
2. Selecting labeled dataset of classes to train the network upon
3. Designing initial network model
4. Training the network
5. Saving the parameters
6. Inference

Fig. 3.8 Deep learning
development process

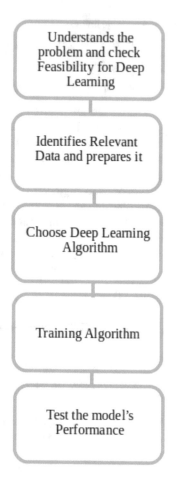

The process can be summarized as shown in Fig. 3.8.

3.7   What Is Deep About Deep Learning?

The traditional neural network consists of at most two layers, and this type of
structure of the *neural network* is not suitable for the computation of larger networks.
Therefore, a neural network having more than 10 or even 100 layers is introduced.

This type of structure is meant for DL. In this, a stack of the layer of neurons is
developed. The lowest layer in the stack is responsible for the collection of raw data
such as images, videos, and text.

Each neuron of the lowest layer will store the information and pass the information further to the next layer of neurons and so on. As the information flows within
the layers of neurons, hidden information of the data is extracted. So, we can

conclude that as the data moves from lowest layer to highest layer (running deep inside the neural network), more abstracted information is collected.

3.8 Data Used for Deep Learning

DL can be applied to any data such as audio, video, text, time series, and images. The features needed within the data are described below:

- The data should be relevant according to the problem statement.
- To perform the proper classification, the dataset should be labeled. In other words, labels have to be applied to the raw dataset manually.
- DL accepts vectors as an input. Therefore, the input dataset should be in the form of vectors and same length. This process is known as data processing.
- Data should be stored in one storage place such as massive file system. If the data is stored in different locations which are not inter-related with each other then, *Data Pipeline* is needed. The development and processing of Data Pipeline is a time-consuming task.

3.9 Difference Between Machine Learning and Deep Learning

DL is fundamentally different from conventional machine learning. In the following example shown in Fig. 3.9, a domain expert would need to spend considerable time in engineering a conventional machine learning system to detect the features that represent a car. With DL, all that is needed is to supply the system with a very large

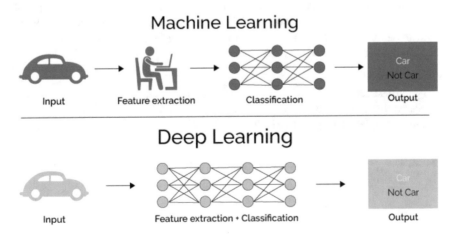

Fig. 3.9 ML vs. DL

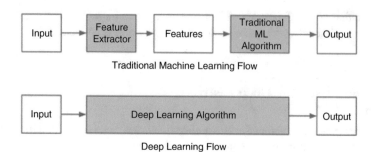

Fig. 3.10 DL vs. ML

Table 3.3 Difference between ML and DL

Machine learning	Deep learning
Traditional ML learning involves manual identification of features and requires much less data	DL algorithms learn high level of features from data; thus, it needs much large data
Dependent on low-end machine	Heavily dependent on high-end machine as large number of matrix multiplication operations which require GPUs and results in the overall increase cost of operation
Divides the tasks into sub-tasks, solves them individually, and finally combines the results. It involves manual feature selection process	Solves problem end to end. Automatically selects the features and assigns the weight
Takes less time to train	Takes longer time to train as there are more parameters
Testing time may increase	Less time to test the data

number of car images, and the system can autonomously learn the features that represent a car.

DL is more powerful and flexible than traditional machine learning. In fact, DL is also a machine learning type but have differences in many other ways. Traditional ML has its own advantages. One has to choose and decide the best for the problem in hand and application. There are four main points to be considered while considering the difference between ML and DL. They are:

- Data dependencies
- Feature selection
- Hardware requirements
- Time complexity (Fig. 3.10 and Table 3.3)

3.10 Why Deep Learning Became Popular Now?

The development of DL is driven by a few forces, and let us discuss this in more detail: DL has found applications since 1990s, but at that time many researchers refused to use it because to make it work perfectly, and for better results one needed a large dataset which can be fed to the network, so that hidden layers can extract every abstract features from it. The age of "Big Data" has made the implementation of DL easier and more effective. As the amount of data increases, the performance of traditional learning algorithms, like logistic regression, does not improve by a whole lot. In fact, it tends to plateau after a certain point. Whereas in the case of deep neural networks, the performance of the model becomes better with more data fed to the model. The amount of data generated every day is staggering—currently estimated to 2.8 quintillion bytes, and this makes DL meaningful and attractive.

DL benefits from the stronger computing power that is available today. The growth of faster computers with larger memory has made the use of DL feasible. Faster CPUs and GPUs provide resources to work on larger data.

By making use of various algorithms, DL can be used to make better business decisions. Researchers are able to perform experiments on a very large scale and new algorithm concepts are evolving with greater accuracy. Because of more accurate results, DL finds increasing use in real-life applications. Many companies like Tesla, Google, Amazon, etc. use DL for their real-world products (Fig. 3.11).

In summary, we can say that following three factors contribute to the popularity of the DL:

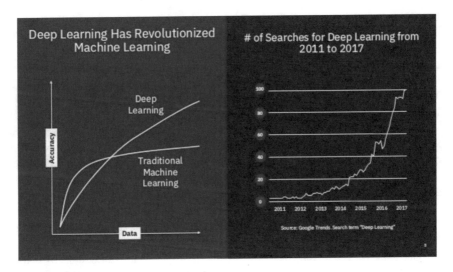

Fig. 3.11 DL popularity

1. Amount of data available
2. Computation time
3. Algorithms

3.11 Should You Always Use Deep Learning Instead of Machine Learning?

The answer is no. This is because DL can be very expensive from a computational point of view. For non-trivial tasks, training a deep neural network will often require processing large amounts of data using clusters of high-end GPUs for many, many hours.

If the problem can be solved using a simpler machine learning algorithm such as Bayesian inference or linear regression, i.e., one that does not require the system to grapple with a complex combination of hierarchical features in the data, then less computational demanding options will be a better choice.

DL may also not be the best choice for making a prediction based on data. For example, if the dataset is small, then sometimes simple linear machine learning models may yield more accurate results—although some machine learning specialists argue a properly trained deep learning neural network can still perform well with small amounts of data.

3.12 Why Is Deep Learning Important?

In today's generation, usage of smartphones and devices have increased drastically. Therefore, more and more images, text, videos, and audios are created every day. It is because deep nets within the DL method can develop a complex hierarchy of concepts. Another point is that when unsupervised data is collected, and machine learning is executed on it, labeling of data has to be performed by the human being. This process is time-consuming and expensive. Therefore, to overcome this problem, DL is introduced as they can identify the particular data.

DL is important for one reason, and one reason only: we have been able to achieve meaningful, useful accuracy on tasks that matter. ML has been used for classification on images and text for decades, but it struggled to cross the threshold. There is a baseline accuracy that algorithms need to have to work in business settings. Computer vision is a great example of a task that DL has transformed into something realistic for business applications. Using DL to classify and label images is not only better than any other traditional algorithms: it is starting to be better than actual human [8].

Facebook has had a great success with identifying faces in photographs by using DL. Software developed by researchers can score 97.25% accuracy regardless of variations in lighting or whether the person in the picture is directly facing the camera [9].

Google is now using DL to manage the energy at their data centers. They were able to cut their energy needs for cooling by 40%. That translates to about 15% improvement in power usage efficiency for the company and hundreds of millions of dollars in savings [10].

3.13   What Are the Drawbacks of Deep Learning?

One of the big drawbacks is the amount of data DL requires to train. For example, Facebook recently announced it had used one billion images to achieve record-breaking performance by an image-recognition system. When the datasets are this large, training systems also require access to vast amounts of distributed computing power. Another issue of DL is the cost of training. Due to the size of datasets and number of training cycles that have to be run, training often requires access to high-powered and expensive computer hardware, typically high-end GPUs or GPU arrays. Whether you are building your own system or renting hardware from a cloud platform, neither option is likely to be cheap. Deep neural networks are also difficult to train, due to what is called the vanishing gradient problem, which can worsen the more layers there are in a neural network. As more layers are added, the vanishing gradient problem can result in it taking an unfeasibly long to train a neural network to a fair level of accuracy.

3.14   Which Deep Learning Software Frameworks Are Available?

Building and deploying DL models proves to be quite a challenging task for scientist and engineers across the industry because of its complexity. Frameworks are tools to ease the building of DL solutions. Frameworks offer a higher level of abstraction and simplify potentially difficult programming task. Thanks to many large tech organizations and open-source initiatives, we now have a plethora of options to choose from. There are a wide range of DL software frameworks as shown in Fig. 3.12, which allow users to design, train, and validate deep neural using a range of different

Fig. 3.12 Popular deep learning frameworks

programming languages. Each framework is built in a different manner for different purposes. Here, we look at some of the top DL frameworks to get a better idea on which framework will be a good fit in solving a particular business problem. These frameworks provide us with reusable code blocks that abstract the logical blocks needed for implementation and also provides several additional modules in developing a DL model.

TensorFlow is one of the most popular DL frameworks. Developed by Google Brain team, allows users to write in Python, Java, C++, and Swift, and that can be used for a wide range of DL tasks such as image and speech recognition, and which executes on a wide range of CPUs, GPUs, and other processors [21]. TensorFlow is an open-source software library for numerical computation using data flow graphs. Nodes in the graph represent mathematical operations, while the graph edges represent the multidimensional data arrays (tensors) communicated between them. It is available on both desktops and cell phones. It has a comprehensive, flexible ecosystem of tools, libraries, and community resources that lets developers easily build and deploy ML-based applications. It is well-documented and has many tutorials and implemented models that are available [11].

Another popular choice, especially for beginners, is PyTorch, a framework that offers the imperative programming model familiar to developers and allows developers to use standard Python statements. It works with deep neural networks ranging from CNNs to RNNs and runs efficiently on GPUs. It employs CUDA along with C/C++ libraries for processing and was made to scale the production for building models and overall flexibility. PyTorch runs on Python, which means that anyone with a basic understanding of Python can get started on building DL models [12].

Among the wide range of other options are Microsoft's Cognitive Toolkit [13], MXNet [14], and Keras [15].

3.15 Classical Problems of Deep Learning Solves

Deep neural networks excel at making predictions based on largely unstructured data. That means they deliver best-in-class performance in areas such as speech and image recognition, where they work with messy data such as recorded speech and photographs.

DL architectures such as deep neural networks, recurrent neural networks, and convolutional neural networks have been applied to fields that include computer vision, speech recognition, natural language processing, audio recognition, social network filtering, bioinformatics, medical image analysis, and material inspection, where they have produced results comparable to and in some cases surpassing human expert performance [3, 16–18].

3.15.1   *Image Classification*

To recognize a human face, first the edges are detected by the DL algorithm to form the first hidden layer. Then, by combining the sides, next shapes are generated as a second hidden layer. After that shapes are combined to create the required human face. In this way, other objects can also be recognized.

Image ==> Edges ==> Face parts ==> Faces ==> Desired face

3.15.2   *Natural Language Processing*

Reviews of movies or videos are gathered together to train them using DL Neural Network for the evaluation of reviews of films.

Audio ==> Low level sound features like (sss, bb) ==> Phonemes == Words == > Sentences

DL neural network plays a major role in knowledge discovery, knowledge application, and last but least knowledge-based prediction.

Areas of usage of DL are listed below.

- Fraud detection
- Customer recommendation
- Self-driving cars
- Analysis of satellite images
- Financial marketing
- Computer vision
- Adding sounds to silent movies
- Automatic handwriting generation
- Image caption generation

3.16   The Future of Deep Learning

Today, there are various neural network architectures optimized for certain types of inputs and tasks. Convolution neural networks are very good at classifying images. Similarly, recurrent neural networks are good at processing sequential data. Both convolution and recurrent neural network models perform supervised learning. Basically, this means they need to be supplied with large amounts of data to learn. In the future, more sophisticated types of AI will use unsupervised learning. A significant amount of research is being devoted to unsupervised and semi-supervised learning technology.

Reinforcement learning is a slightly different paradigm to DL in which an agent learns by trial and error in a simulated environment solely from rewards and punishments. DL extensions into this domain are referred to as deep reinforcement learning (DRL). There has been considerable progress in this field, as demonstrated by DRL programs beating humans in the ancient game of GO.

Designing neural network architectures to solve problems becomes more complex with many hyperparameters to tune and many loss functions to choose from to optimize. There has been a lot of research activity exploring neural network architectures to operate autonomously. Learning to learn, also known as meta-learning or Auto ML is a step in this direction.

Current artificial neural networks were based on 1950s understanding of how human brains process information. Neuroscience has made considerable progress since then, and DL architectures have become so sophisticated that they seem to exhibit structures such as grid cells, which are present in biological neural brains used for navigation. Both neuroscience and DL can benefit each other from cross-pollination of ideas.

3.17 Points to Ponder

1. What are the differences between machine learning (ML) and deep learning (DL)?
2. What are the advantages of DL?
3. Are there drawbacks of DL?

References

1. Dechter, R. (1986). *Learning while searching in constraint-satisfaction problems*. Los Angeles: University of California, Computer Science Department, Cognitive Systems Laboratory.
2. Krizhevsky, A., Sutskever, I., & Hinton, G. E. (2012). Imagenet classification with deep convolution neural networks. *Advances in Neural Information Processing Systems*, 1097–1105.
3. https://en.wikipedia.org/wiki/Deep_learning.
4. Graves, A., Liwicki, M., Fernandez, S., Bertolami, R., Bunke, H., & Schmidhuber, J. (2009). A novel connectionist system for improved unconstrained handwriting recognition. *IEEE Transactions on Pattern Analysis and Machine Intelligence, 31*(5), 855–868.
5. Li, X., & Wu, X. (2014). Constructing long short-term memory based deep recurrent neural networks for large vocabulary speech recognition.
6. Bengio, Y., LeCun, Y., & Hinton, G. (2015). Deep learning. *Nature, 521*(7553), 436–444.
7. Goodman, I., Bengio, Y., & Courville, A. (2016). Deep Learning (Adaptive Computation and Machine Learning Series), MIT Press
8. https://algorithmia.com/blog/introduction-to-deep-learning.
9. https://research.fb.com/blog/2016/08/learning-to-segment/.
10. https://www.vox.com/2016/7/19/12231776/google-energy-deepmind-ai-data-centers.
11. https://www.tensorflow.org/.
12. https://pytorch.org/.
13. https://docs.microsoft.com/en-us/cognitive-toolkit/.
14. https://mxnet.apache.org/.
15. https://keras.io/.

16. https://machinelearningmastery.com/inspirational-applications-deep-learning/.
17. https://medium.com/breathe-publication/top-15-deep-learning-applications-that-will-rule-the-world-in-2018-and-beyond-7c6130c43b01.
18. http://www.yaronhadad.com/deep-learning-most-amazing-applications/.
19. https://en.wikipedia.org/wiki/Recurrent_neural_network.
20. Dupond, S. (2019). A thorough review on the current advance of neural network structures. *Annual Reviews in Control, 14*, 200–230.
21. https://www.nvidia.com/en-us/geforce/gaming-laptops/20-series/.

Chapter 4
Cloud Computing Concepts

4.1 Roots of Cloud Computing

Although cloud computing has been in vogue for over a decade, its roots go back nearly half a century [1]. With the advent of mainframe computers, at the end of World War II, many users were required to share large computing machines to defray the cost of buying and maintaining them. Then in the subsequent decades, computing pendulum swung to put more control in the hands of local personal computer users. This led to the development of many games and smaller applications such as word processing and accounting applications. However, a need was felt to share this data and applications among a group of users, such as within an enterprise. Thus, client–server computing concepts were invented, giving rise a large central computer connected to weaker client machines, also known as thin clients. An enabling technology, i.e., computer-to-computer communication, made the client–server implementations economically viable. Then came hand-held devices, such as tablets and smartphones, giving computational power back to the local users for collecting, generating, and sharing data. However, this phase lasted less than a decade as a need was felt to store this data in the backend servers for ease of storage and retrievals. These phases of evolution are numbered and shown in Fig. 4.1. Currently, we are in the middle phase of computing pendulum oscillations, where both the local and remote computing are considered important.

While this pendulum keeps swinging back and forth, the data center end of computing quietly evolved from single, large mainframe machines to a collection of servers, connected by various networking protocols. These appeared to give the impression of a single large machine, available to the users in the middle and left side of the computing spectrum, as shown in Fig. 4.1. Before the popularity of cloud computing term, other terms were in vogue to describe similar services, such as grid computing to solve large problems using parallelized solutions, e.g., in a server farm. Another precursor to cloud's promise of anytime, anywhere access to IT resources was utility computing. In utility model, computing resources were offered as a

P. Gupta, N. K. Sehgal, *Introduction to Machine Learning in the Cloud with Python*, https://doi.org/10.1007/978-3-030-71270-9_4

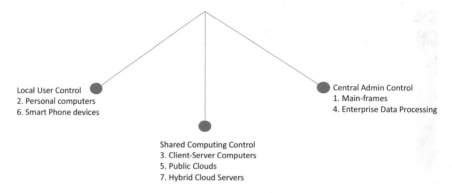

Local User Control
2. Personal computers
6. Smart Phone devices

Central Admin Control
1. Main-frames
4. Enterprise Data Processing

Shared Computing Control
3. Client-Server Computers
5. Public Clouds
7. Hybrid Cloud Servers

Fig. 4.1 Three stages of computing pendulum swings

metered service. An additional feature of cloud is elasticity, which allows rapid scaling of IT resources to meet an application's demand, without having to forecast or provision for it in advance.

4.2   Key Characteristics of Cloud Computing

National Institute for Standards and Technology (NIST) has defined cloud computing [2] as a model for enabling ubiquitous, convenient, on-demand network access to a shared pool of configurable computing resources. These resources include networks, servers, storage, applications, and services.

Cloud computing has been driven by economic considerations to share the cost of maintaining resources in centralized data centers. These resources are made available by remote access to many users, similar to other public utilities such as electrical grid or water supply. The users pay for what they use and may pay another fee for having the access availability.

NIST characterizes five essential features of any cloud computing service:

1. *On-demand self-service*: A consumer can utilize computing capabilities, such as server time and network storage, as needed automatically without requiring human interaction with each service provider.
2. *Broad network access*: Capabilities are available over the network and accessed through standard mechanisms that promote use by heterogeneous thin or thick client platforms (e.g., mobile phones, tablets, laptops, and workstations).
3. *Resource pooling*: The provider's computing resources are pooled to simultaneously serve multiple consumers using a multi-tenant model, with different physical and virtual resources dynamically assigned and reassigned according to consumer demand. There is a sense of location transparency in that a customer does not generally care about the exact location of the provided resources but may be able to specify location at a higher level of abstraction (e.g., country, state, or

datacenter). Examples of resources include storage, processing, memory, and network bandwidth.

4. *Rapid elasticity*: Capabilities can be elastically provisioned and released, in some cases automatically, to scale rapidly scale-up or -down commensurate with demand as needed. To the consumer, the capabilities available for provisioning often appear to be unlimited and can be appropriated as much as required.

5. *Measured service*: Cloud systems automatically control and optimize resource use by leveraging a metering capability at some level of abstraction appropriate to the type of service (e.g., storage, processing, bandwidth, and active user accounts). Resource usage can be monitored, controlled, and reported, providing transparency for both the provider and consumer of the utilized service.

A cloud computing service is available to the users at three different levels, also known as the service models:

1. *Software as a Service (SaaS)*: The capability provided to the consumer is to use the provider's applications running on a cloud infrastructure. The applications are accessible from various client devices through either a thin client interface, such as a web browser (e.g., web-based email), or a program interface. The consumer is relieved from the concerns of management or control of the underlying cloud infrastructure including network, servers, operating systems, storage, or even individual application capabilities, with the possible exception of limited user-specific application configuration settings. Such a service model is used by Salesforce.com, which offers CRM (customer relationship management) tools to its customers, using Amazon's public cloud data centers.

2. *Platform as a Service (PaaS)*: The capability provided to the consumer is to deploy onto the cloud infrastructure consumer-created or acquired applications created using programming languages, libraries, services, and tools supported by the provider. The consumer does not manage or control the underlying cloud infrastructure including network, servers, operating systems, or storage but has control over the deployed applications and possibly configuration settings for the application-hosting environment. Such a service model is used by Google's Cloud Platforms (GCP), offering AI/ML (artificial intelligence and machine learning) tools using its public facing data centers.

3. *Infrastructure as a Service (IaaS)*: The capability provided to the consumer is to provision processing, storage, networks, and other fundamental computing resources where the consumer is able to deploy and run arbitrary software, which can include operating systems and applications. The consumer does not manage or control the underlying cloud infrastructure but has control over operating systems, storage, and deployed applications; and possibly limited control of select networking components (e.g., host firewalls). Such a service model is used by AWS (Amazon Web Services) renting various Linux- or Windows-based server platforms on an hourly basis to its customers.

Additional service models are emerging, such as DaaS (Data as a Service) or AI as a Service, but these are yet to become prevalent as the three listed above. AI/ML can be offered at any of the above three layers of abstraction depending on the nature of usage. Meanwhile, cloud service locations have evolved from large, localized data centers to various deployment models, categorized by NIST as follows:

1. *Private cloud*: The cloud infrastructure is provisioned for exclusive use by a single organization comprising multiple consumers (e.g., business units). It may be owned, managed, and operated by the organization, a third party, or some combination of them, and it may exist on or off premises.
2. *Community cloud*: The cloud infrastructure is provisioned for exclusive use by a specific community of consumers from organizations that have shared concerns (e.g., mission, security requirements, policy, and compliance considerations). It may be owned, managed, and operated by one or more of the organizations in the community, a third party, or some combination of them, and it may exist on or off premises.
3. *Public cloud*: The cloud infrastructure is provisioned for open use by the general public. It may be owned, managed, and operated by a business, academic, or government organization, or some combination of them. It exists on the premises of the cloud provider.
4. *Hybrid cloud*: The cloud infrastructure is a composition of two or more distinct cloud infrastructures (private, community, or public) that remain unique entities, but are bound together by standardized or proprietary technology that enables data and application portability (e.g., cloud bursting for load balancing between clouds).

4.3   Various Cloud Stakeholders

As one can imagine, the popularity and spread of cloud computing has resulted in a value chain that spans large geographies and involves many players. Just like any supply chain, if we start with the providers that is the data center owners and managers, traversing all the way to the end-users, following list emerges [3]:

1. *Cloud provider*: An organization or entity that provides cloud services to cloud consumers.
2. *Cloud carrier*: It works as glue in cloud ecosystem between cloud consumers and cloud service providers (CSP). CSPs use it for connectivity and transport of cloud services to consumers.
3. *Cloud broker*: Often, cloud brokers are responsible for managing delivery, performance, and quality of cloud services to the cloud consumers.
4. *Cloud consumer*: A cloud consumer user services from CSPs. The consumer can be an end user, organization, or set of organizations having common regulatory constraints; performs a security and risk assessment for each use case of cloud migrations and deployments.

5. *Cloud auditor or cloud aware auditors*: Cloud aware auditors conduct third-party assessment of cloud services, information system operations, performance, and security of the cloud implementation based on existing rules and regulations.

All of the above stakeholders share some common goals, such as information security and operational efficiency but also have conflicting interests. A cloud consumer wishes to pay the minimum amount of money but wants to get maximum computing to meet his or her needs. In contrast, a cloud provider wants to maximize profits by placing more customer workloads [4] on the same share hardware, potentially causing latency for the cloud consumers. These often get resolved by the terms listed in an SLA (service level agreement). However, most SLAs are at a higher level of abstraction listing primarily the uptime for servers and network, but not necessarily their performance levels. For mission critical applications, customers often resort to hybrid clouds [5] or multi-vendor solutions [6], but migration of data needs to be addressed.

4.4 Pain Points in Cloud Computing

Figure 4.2 outlines pain points for various stakeholders, starting with cloud users who must specify the type and range of services they need in advance. If future requests exceed forecast, then level of service will depend on the elasticity of providers' capacity as cloud hardware is shared among many tenants. So, the level of service may deteriorate for all the users on a given server, e.g., if one of the users starts to do excessive I/O operations, it will slow down I/O requests from other users. This is referred to as a noisy neighbor virtual machine (VM) problem such as caused by placing a streaming media task next to the database-accessing job on the same server. A cloud user may care only for his or her own QoS, while a cloud service provider (CSP) makes the best effort to satisfy all users. IT managers need to ensure that hardware is being fully utilized, and the data center is receiving adequate power and cooling, etc. There is no silver bullet to solve all cloud stakeholders' problems. However, cloud usage still keeps on growing due to economic considerations.

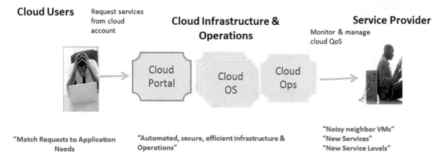

Fig. 4.2 Cloud computing pain points

4.5 AI and ML in Cloud

As we learned in the previous chapters that artificial intelligence (AI) and machine learning (ML) refer to tools and techniques to learn from training data [7]. The strategy is to detect patterns and predict future behaviors. One of the goals of AI- and ML-based solutions is to perform specific tasks without using explicit instructions. A set of observations in the form of input–output pairs are used as training set(s). These sample datasets are used to build mathematical models and determine future decision parameters for ML algorithms. Since the size of data to train a model can be very large, and computations very complex, it is imperative that elastic resources offered by a public cloud are very well-suited to meet AI and ML needs. This allows the cloud users to pay only for the compute services they need without having to provision for them in advance.

Specific implementation details of ML techniques such as neural networks, DTs, and association rules are discussed elsewhere in this book. Here we will examine the set of services that are currently available in a public cloud. For example, Sagemaker [8] is a fully managed service offered by Amazon that works in three steps, as shown in Fig. 4.3.

1. *Sample data generation*: It involves fetching the data, cleaning it for consistency, and transforming it to improve ML performance. Example of cleaning includes edits to make sure that same items are referred to by the same names. For example, US, USA, United States, and United States of America refer to a country with the same name attribute. Transformation example includes combining multiple attributes into a single one for making decisions, namely temperature

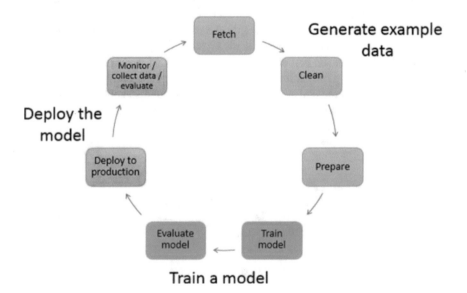

Fig. 4.3 Typical workflow for creating an ML model

and humidity can be used as a two-tuple to determine if air-conditioning should be turned on.

2. *Building and evaluating a model*: Once the input data is ready, then one of the available algorithms can be used for model training. Then accuracy of its inferences is drawn to see if the model is acceptable or not.
3. *Deploying the model*: This is the last and final step which is to use the model for actual data going beyond the initial training set. ML is a continuous improvement cycle, so inferences are often checked to minimize drifts over time. If output results are not desirable, then the model needs to be retrained with a new dataset.

Note that the amount of compute needed for training is more than what is required for inference applications. Due to this elastic nature of computational resources required during different phases of AI and ML lifecycle, it is desirable to have flexibility that only a public cloud can offer. In a private cloud or enterprise-based data center, the peak levels of demand will need to be understood in advance. Then servers, storage, and networking equipment must be procured in advance and provisioned. Even then most of it may be underutilized during the non-peak hours, raising the overall cost of an AI or ML solution. Thus, a public cloud is better suited to perform such activities.

Some stakeholders may cite confidentiality as the reason to avoid a public cloud. Such objections can be overcome with a combination of data encryption and hybrid cloud solutions, which we shall explore in later chapters. Furthermore, cloud users may not always own the data they need for their AI and ML tasks. An example is healthcare data owned by hospitals, which are willing to share it with medical researchers after some anonymization to help with discovery of new drugs.

Many online businesses are using cloud analytics to better serve their existing customers and attract new customers. An example is Amazon's bookstore suggesting additional titles when a book is purchased, based on what other readers are reading, or Google sending targeted customized advertisement to its search or Gmail users based on their search patterns or social media activities. Some example uses of data analytics in the cloud are as follows:

1. *Social media*: Billions of users are active on applications such as Facebook, Instagram, and Twitter, sharing their stories, opinions, and preferences. This is heaven for marketers to identify potential customers as well as looking for what others are saying about their products or services online. By searching and linking activities across different online sites, it has become easier than ever before to construct a customer's profile, even without meeting that person.
2. *Tracking products*: An online business can track their inventory across warehouses and ship items to customers from the nearest location to minimize shipping time and costs. Similarly, new products can be ordered to replenish supplies, and returns can be tracked in an automated way.
3. *Tracking preferences*: Online movie and song companies log what each user watches or listens to. Then this information is used to recommend other movies or songs along similar themes, to keep the user interested. Internet has become a

battleground for eyeballs and mindshare to keep user engaged, so more services
and advertisements can be served in a relevant manner.

4. *Keeping records*: Cloud enables real-time recording and sharing of data regardless of location. An example is an online retailer notifying customers when goods
are delivered at their home. Furthermore, a facility is offered to alert and buy the
next round of supplies at home just before the previous batch finishes. This data is
stored in cloud to track patterns across regions and seasons, so business can stock
their goods in an efficient manner.

An advantage of data analytics in cloud is that entire datasets can be used instead
of smaller statistical samples that may not represent the heterogeneity of a big data
set. This helps to eliminate guesswork and enables identification of data patterns to
minimize uncertainty.

Another advantage of conducting AI and ML in a public cloud is the ability to
combine data from multiple sources. We revisit the case of a hospital providing its
patient data after some anonymization. That may be necessary but not sufficient for
the medical researchers to discover new drugs, say for a deadly disease such as
cancer. They also need drug efficacy data from various medicine makers, which may
be even competing with each other for future trials and business success. In this case,
none of the parties has all the ingredients of a possible solution. However, they all
need to come together in a place that offers them a level playing field. Public cloud
can provide such an opportunity, provided the data security is assured and a mutual
access benefits all parties. This is the newest trend and refers to as privacy preserving
machine learning (PPML) analytics, which we will study in the later chapters.

4.6 Expanding Cloud Reach

With the advent of intelligent end-point devices, such as Internet-enabled smart TVs,
and home surveillance cameras, came the birth of Internet of Things (IoT). It refers
to computing devices embedded in everyday objects, enabling them to send and
receive data. The definition of IoT has evolved due to the convergence of multiple
technologies, namely embedded systems with commodity sensors to support real-
time analytics and machine learning [9]. Since most of these devices in the field are
connected to cloud data centers for storage and processing, IoT devices represents an
expansion of the cloud as defined by NIST.

In the consumer market, IoT technology refers to products and services building
up the idea of a "smart home," enabling devices and appliances that can be
controlled remotely. IoT also has reached other market segments such as industrial
applications, medical facilities, transportation, etc. A seamless integration of various
manufacturing devices equipped with sensing, identification, processing, communication, actuation, and networking capabilities gives rise to a smart cyber-physical
space (CPS). This has helped to create many new businesses and market opportunities. Industrial IoT (IIoT) in manufacturing is already generating so much business

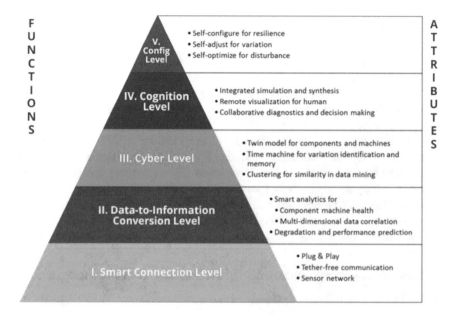

Fig. 4.4 Architecture of cyber-physical system-enabled manufacturing system

value leading to a Fourth Industrial Revolution, which is estimated to generate $12 trillion of global GDP by 2030 [10].

CPS is the core technology of industrial big data [11] and represents an interface between humans and the cyber world. Such system can be designed by following the 5C (Connection, Conversion, Cyber, Cognition, Configuration) architecture, as shown in Fig. 4.4. However, a growth in IoT has also contributed to a growing number of serious concerns in areas of privacy and security. Weak authentication, unencrypted messages sent between devices, and poor handling of security updates can compromise IoT devices. Most of these concerns are similar to those for the cloud servers, workstations, and smartphones, represented by confidentiality, integrity, and availability (CIA). In addition, it is possible to use internet connected appliances to "spy on people in their own homes" by using smart televisions or kitchen appliances. There have been instances of botnet attacks by hijacking many home security cameras to inundate Cloud servers [12].

Machine learning (ML) offers a possibility to discover and plug-in vulnerabilities. At the same time, bad actors may also use ML to identity some vulnerabilities and attack. ML has been very successful with complex tasks such as image recognition and natural language processing [13]. ML is now beginning foray into cybersecurity. However, security problem is more complex as we have human attackers striving to compromise the security of a system, so the nature of problem keeps changing dynamically. Let us start by examining some newly emerging applications that traditional security algorithms cannot address efficiently.

4.7   Future Trends

As cloud workloads are increasingly gravitating toward AI and ML type of compute intensive algorithms, some customers are feeling a need to accelerate these tasks. One way to achieve this is by using graphics processing units (GPUs) because many of these proffer parallel processing capabilities, which are useful in image processing and graph-based searches.

In a traditional cloud, computations are done by multi-core CPUs and many core GPUs, with a suitable allocation of resources to match the algorithmic needs. Lately, GPUs have evolved from graphics-specifics to general purpose compute devices, initiating a new era of computing known as GPGPUs. New software stacks have been developed to harness the full power of GPGPUs with virtualization with appropriate programming models [14]. An example is shown in Fig. 4.5, with multiple layers of hardware and software stacks, and mapping between them.

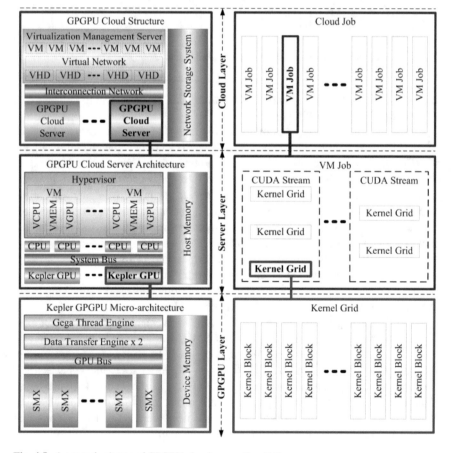

Fig. 4.5   A general scheme of GPGPU cloud computing [14]

There are three layers for mapping different computation tasks to hardware components, namely:

1. *GPU layer*: This contains the hardware resources, including data transfer engine (DTE), a giga thread engine (GTE) and memory shared between them. All components connect through an interconnected network, such as a GPU bus.
2. *Server layer*: It contains a hypervisor to manage all physical devices, and a virtual machine to invoke kernel level tasks. This layer is responsible for scheduling, memory loading, and preparing for program execution.
3. *Cloud layer*: This is the uppermost layer, containing a virtualization management environment. It maintains a pool of virtual machines (VMs), a cluster of GPGPU servers holding the resources for the VM instances, and a network storage system managing a pool of virtual hard disks (VHDs). All cloud servers in this layer are connected through a low latency network, such as InfiniBand (IB).

Note that a cloud job may contain several parallel VMs, with each VM carrying out an atomic task. A cloud user is able to specify mapping of a VM to the number of virtual CPUs, virtual GPUs, virtual memory and VHD capacity and the virtual network bandwidth. Entire computation can be distributed in a hierarchical or flat manner. Future GPGPU computing paradigms have four key features:

1. *Virtualization*: This enables higher utilization of multiple tasks in a cloud.
2. *Hierarchical*: This enables mapping of GPGPU tasks to the hardware structure, to achieve a balance of resource capacity and task flexibility.
3. *Scalable*: This allows a task with deterministic computation complexity to be arranged as a group of VMs on multiple VPGPUs, or as one VM on a single VGPU. Resulting difference will be in the run time, with more physical resources leading to a faster execution time.
4. *Diverse*: This allows a mix of workloads to be mapped in parallel to different types of hardware, for better execution performance.

The last feature is now leading to a next step in evolution to other types of specialized hardware accelerators, such as for tensor processing units (TPUs) for tensor flow execution. We will revisit these in one of the later chapters.

4.8 Summary

Cloud computing has been evolving and expanding over the past decade, to encompass Internet-connected devices in the field. This offers new business opportunities and additional security challenges. Former is driven by automation, to offer 24×7 monitoring with surveillance cameras, improving industrial productivity and scalability never seen before in the human history. However, addition of edge computing and IoT also expands the attack surface and threat models. This enable hackers to operate at a massive scale by remotely hijacking IoT devices, and launch remote Denial of Service (DoS) attacks on centralized servers in a cloud datacenter. Hence, there is a need for new machine learning-based techniques to improve cloud computing security that we shall study in the later chapters.

4.9 Points to Ponder

1. If a cloud service provider wants to offer ML at IaaS layer, what will be the features of such a service?
2. If a cloud service provider wants to offer ML at PaaS layer, what will be the features of such a service?
3. If a cloud service provider wants to offer ML at SaaS layer, what will be the features of such a service?

References

1. Sehgal, N. K., Bhatt Pramod Chandra, P., & Acken, J. M. (2019). *Cloud computing with security*. New York: Springer. https://rd.springer.com/book/10.1007/978-3-030-24612-9.
2. https://csrc.nist.gov/publications/detail/sp/800-145/final.
3. https://clean-Clouds.com/2013/12/18/stakeholders-in-Cloud-computing-environment/.
4. Mulia, W. D., Sehgal, N., Sohoni, S., Acken, J. M., Stanberry, C. L., & Fritz, D. J. (2013). Cloud workload characterization. *IEIE Technical Review*, 382–397. https://www.tandfonline.com/doi/abs/10.4103/0256-4602.123121.
5. https://www.redhat.com/en/topics/Cloud-computing/what-is-hybrid-Cloud.
6. https://www.cio.com/article/3183504/why-your-Cloud-strategy-should-include-multiple-vendors.html.
7. https://en.wikipedia.org/wiki/Machine_learning.
8. https://docs.aws.amazon.com/sagemaker/latest/dg/how-it-works.html.
9. https://en.wikipedia.org/wiki/Internet_of_things.
10. Daugherty, P., Negm, W., Banerjee, P., & Alter, A. Driving unconventional growth through the industrial internet of things (PDF). Accenture. Retrieved 17 Mar 2016.
11. https://www.accenture.com/us-en/_acnmedia/accenture/next-gen/reassembling-industry/pdf/accenture-industrial-internet-changing-competitive-landscape-industries.pdf.
12. https://www.csoonline.com/article/3258748/the-mirai-botnet-explained-how-teen-scammers-and-cctv-cameras-almost-brought-down-the-internet.html.
13. https://towardsdatascience.com/machine-learning-for-cybersecurity-101-7822b802790b.
14. Hu, L., Che, X., & Xie, Z. (2013). GPGPU Cloud: A paradigm for general purpose computing. *Tsinghua Science and Technology, 18*(1), 22–33. ISSN 1007-2014. https://ieeexplore.ieee.org/stamp/stamp.jsp?arnumber=6449404.

Part II
Practices

Chapter 5
Practical Aspects in Machine Learning

5.1 Preprocessing Data

Data is truly considered a major resource in today's world. As per the World Economic Forum, by 2025, 463 exabytes (10^{18}) of data will be generated globally per day. One question that arises: Is all of this data good enough to be used by machine learning algorithms? How do we decide that? In this chapter, we will explore the topic of data preprocessing, i.e., transforming the raw data so that it becomes suitable for the machine learning algorithms. Data cleaning and preparation is a critical first step in all machine learning projects. Although data scientists spend considerable amount of time tinkering with algorithms and machine learning models, the reality is that data scientists spend most of their time 70–80% in data preparation or data preprocessing. Data preprocessing is an integral step in machine learning as the quality of data and the useful information that can be derived from it directly affects the ability of ML model to learn.

When we talk about data, we envisage large datasets as having a huge number of rows and columns as in a large database. While it is a likely scenario, it is not always the case. The data could be in so many different forms such as structured tables, unstructured text data containing images, audios, and video files in a variety of format. Since machines can interpret strings formed with 1's and 0's. Data needs to be transformed or encoded to bring it to such a state that the machine can easily parse it to interpret for use in machine learning algorithms. This is what is achieved during preprocessing stage [1].

Data preprocessing is a technique that is used to convert the raw data into a clean data set. Initially the data is gathered from different sources in raw format that is not suitable for analysis. Real-world data is often incomplete, inconsistent, and/or lacking in certain details to reveal behavior or trends. Also, it is likely to contain many errors. Data preprocessing is a proven method of resolving such issues. For achieving good results in ML projects, data has to be in a proper format. Some specified ML algorithms need information in a specified format. For example, most

P. Gupta, N. K. Sehgal, *Introduction to Machine Learning in the Cloud with Python*, https://doi.org/10.1007/978-3-030-71270-9_5

of the ML algorithms do not support missing or null values. Therefore, to execute algorithms, the null/missing values have to be managed. Another aspect is that data should be formatted in such a way that more than one ML algorithm can be executed on a dataset, and the one performing better is chosen. We will discuss some important aspects of data processing in this chapter.

5.2   Challenges in Data Preparation

Data preparation refers to transforming raw data into a form that is better suited for ML modeling. As such, choice and configuration of data preparation applied to the raw data may be treated as another hyperparameter of the modeling pipeline to be tuned. The framing of data preparation can be overwhelming to beginners given the large number and variety of data. The solution to this problem is to think about data preparation techniques in a systematic way.

5.3   When to Use Data Preprocessing?

In the real world, most of the data are noisy. It often contains errors making it difficult to use and analyze. Sometimes data is unstructured and needs to be transformed into a structured form before we can use it for analysis and modeling. Transforming the unstructured data into structured data requires data preprocessing.

Let us look at the objectives of data preprocessing:

- Recognize the importance of data preparation
- Identify the meaning and aspects of feature engineering
- Deal with missing values and outliers
- Standardize data features with feature scaling
- Analyze the data: summary statistics and visualizations
- Does the data need to be aggregated?
- Dimensionality reduction with Principal Component Analysis (PCA) ought to be explicit

5.4   Framework for Data Preparation Techniques

There are a number of different approaches and techniques for data preparation that could be used during machine learning process. As stated earlier, data preparation [4] can be treated as another hyper-parameter to tune as part of the modeling pipeline. Data preparation allows one to ensure that effective techniques are explored and not skipped or ignored. This can be achieved using a suitable framework to organize data

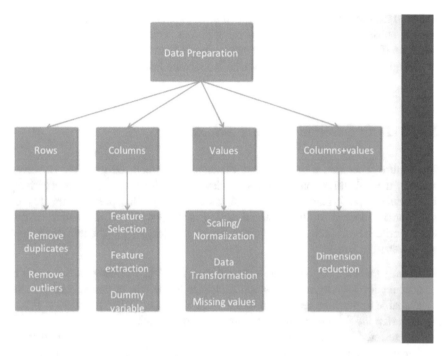

Fig. 5.1 Data preparation framework

preparation techniques that consider their effect on the raw dataset. For example, structured data, such as data that might be stored in a CSV file, consists of rows, columns, and values. Various techniques can be considered that operate at each of these values.

- Data preparation for rows
- Data preparation for columns
- Data preparation for values

Data preparation for rows may use techniques that add or remove duplicate or missing data from the dataset. Similarly, data preparation for columns may be techniques that add or remove columns (feature selection/feature extraction) from the dataset. Fig. 5.1 shows the framework for data preparation [2, 3].

Next we will discuss various phases during data preparation.

5.4.1 Data Preparation

Data preparation is the process of transforming raw data into meaningful features for a data mining task, which act as input for ML algorithms and help in improving the overall ML model performance. Generally, it starts from an initial act of measured data and lead to derived features intended to be explanatory and essential, simplifying the subsequent learning and modeling phases. Data preparation involves data selection (feature selection), data preprocessing, data transformation, etc. Below, we give a quick brief about data preparation in machine learning is mentioned below.

5.4.2 Data Selection (Feature Selection)

The feature generation allows to transform data into synthetic information. However, some of these features could be either irrelevant or redundant and could negatively influence the performance of the trading activity. For this purpose, the feature selection is adopted by this protocol, and it is done with the application of two cutting criteria: correlation coefficient and multicollinearity [5]. The following steps are involved in data selection:

- Usually there is a huge volume and vast variety in data. Besides this, one needs to account for velocity in data which translates to the rate at which data is made available to machine learning model.
- This step involves selecting only subset of the available data.
- The selected sample should be fairly accurate representation of the entire dataset.
- Some data can be derived or simulated from the available data if required.
- Data not relevant can be excluded.

5.4.3 Data Preprocessing

Data preprocessing involves the following steps:

- Format the data to make it suitable for ML.
- Clean the data to remove incomplete variables.
- Sample the data to reduce training time and memory requirements.

5.4.4 Data Cleaning

Data cleaning is the process of identifying and correcting (or removing) incomplete, improper and inaccurate data. The aim is to address what are referred to as data quality issues, which negatively affect the quality of model and compromise the analysis process and results. There are several types of data quality issues, including missing values, duplicate data, outliers, inconsistent or invalid data. We will discuss these issues and how to handle them in the chapter later.

5.4.5   Insufficient Data

The amount of data required for ML algorithms can vary from thousands to millions, depending upon the complexity of the problem and the chosen algorithm. Selecting the right size of the sample is a key step in data preparation. Samples that are too large or too small might give skewed results. Smaller data set cause sampling noise since the algorithm gets trained on non-representative data. For example, checking voter sentiments for a very small subset of voters. Larger samples work well as long as there is no sampling bias. For example, sampling bias would occur when checking voter sentiment only for technology savvy subset of voters, while ignoring others.

5.4.6   Non-representative Data

The sample selected should be a fair representation of the entire data. A non-representative data might train an algorithm such that it will not generalize well on new data or unseen data.

5.4.7   Substandard Data

Outliers, errors, and noise can be eliminated to get a better fit of the model. Missing features such as salary of 10% of audience may be ignored completely, or an average value can be assumed for the missing value. While taking any action one has to be very careful as bias can be easily cropped.

5.4.8   Data Transformation

The selected and preprocessed data is transformed using one or more of the following methods:

- Scaling: It involves selecting the right feature scaling for the selected and preprocessed data discussed later in the chapter.
- Aggregation: This is to collate a bunch of data features into a single one.

5.4.9   Handling Missing Values

In real-world data, there are some instances where a particular element is absent because of various reasons, such as corrupted data, failure to load the information, or incomplete extraction. Handling the missing values is one of the challenges faced by analysts. Making the right decision on how to handle it generates robust data models. If the missing values are not handled properly, then one may end up drawing an inaccurate inference about the data. There are various methods to deal with missing values. Let us look at some of the ways to handle the missing values.

1. Deleting Rows/Columns

   This method is commonly used to handle the null value. In this method, we either delete a particular row and a particular column if it has more than 60–70% of missing values. When there are lots of missing values it implies that particular feature/column may not be important. This method is advised only when there are enough observations. One has to make sure that removing the data will not add bias. Moreover, removing the data will lead to loss of information that may not give the expected results.

2. Replacing with Mean/Median/Mode

   Missing values are replaced with mean, median or mode of the feature. This is an approximation and may add variance. But loss of the data can be negated by this method that yields better results compared to removal of data. This is a statistical approach to deal with handling the missing values.

3. Assigning a Unique Category

   A category variable has a definite number of categories, such as email (spam or no spam). Since they have a definite number of classes, we can assign another class for the missing values. We can replace missing values with a new category, say, "Unknown."

4. Predicting the Missing Values

   We can predict the missing values with the help of some modeling methods like linear regression, etc. This method may result in better accuracy, unless a missing value is expected to have a very high variance. One can experiment with different algorithms and check, which gives better accuracy instead of sticking to a single algorithm.

Note: By imputing the missing values, one has to be careful as it may introduce bias.

5.5 Modification of Categorical or Text Values to Numerical Values

Machine learning algorithms cannot work with categorical data directly. Categorical data needs to be encoded to numerical forms. Data Preprocessing in machine learning requires values of the data in numerical form. Therefore, we have to encode text values in the columns of datasets into numerical form. For example, the LabelEncoder() class in sklearn library can be used to transform the categorical or string variable into the Numerical Values. There is one problem with labelEncoder()

as the equation in the model may introduce order in categories. This can be prevented by creating dummy variables. Dummy variables take the values 0 or 1 to indicate the absence or presence of some categorical effect that may be expected to shift the outcome. Instead of having one column, we will have as many as additional columns as number of categories in the feature/column. A one hot encoding is a representation of categorical variables as binary vectors. This first requires that the categorical values be mapped to integer values.

Example:

Assume we have a sequence with the values 'male' and 'female'. We can assign 'male' an integer value of 1 and 'female' the integer value of 0. As long as we always assign numbers to these labels, this is called integer encoding. Next, we can create a binary vector to represent each value. The vector will have a length of 2 for the possible integer values. The "male" label encoded as a 1 will be represented with a binary vector [1,0] where the zeroth index is marked with a value of 1. In turn, the "female" label encoded as a 0 will be represented with a binary vector [0,1].

If we had the sequence

'male', female', 'male', 'male'

we could represent it with the integer encoding

1, 0, 1,1

and the one hot encoding of:

[1,0]

[0,1]

[1,0]

[1,0]

One Hot Encode with scikit-learn

An example sequence is as follows and consists of two labels: 'male' and 'female'

data = ['male', 'female', 'female', 'female', 'male', 'male', 'female', 'male', 'female', 'female']

Here we will use the encoders from the scikit-library. Specifically, the "labelEncoder" of creating an integer encoding of labels and the "OneHotEncoder" for creating a one hot encoding of integer encoded values.

```
from numpy import array
from numpy import argmax
from sklearn.preprocessing import LabelEncoder
from sklearn.preprocessing import OneHotEncoder

# define example

data = ['male', 'female', 'female', 'female', 'male', 'male', 'female', 'male',
'female', 'female']
values = array(data)
print(values)
```

```
output:
['male' 'female' 'female' 'female' 'male' 'male' 'female' 'male'
'female' 'female']
```

```
# integer encode
label_encoder = LabelEncoder()
integer_encoded = label_encoder.fit_transform(values)
print(integer_encoded)
```

```
[1 0 0 0 1 1 0 1 0 0]
```

```
# binary encode
onehot_encoder = OneHotEncoder(sparse=False)
integer_encoded = integer_encoded.reshape(len(integer_encoded), 1)
onehot_encoded = onehot_encoder.fit_transform(integer_encoded)
print(onehot_encoded)
```

```
[[0. 1.] [1. 0.] [1. 0.] [1. 0.] [0. 1.] [0. 1.] [1. 0.] [0. 1.] [1. 0.]
[1. 0.]]
```

Running the example first prints the sequence of labels. This is followed by the integer encoding of the labels and finally the one hot encoding.

5.6   Feature Scaling

When data has attributes with varying scales, it may be helpful to rescale. Many machine learning algorithms can benefit from rescaling the attributes to all have the same scale.

Feature scaling is an important step in the data transformation stage of the data preparation process. Feature scaling is the method to limit the range of variables so that they can be compared on common grounds.. Feature scaling is a method for standardization of independent features. It means to adjust values of numeric features measured on different scales to a notionally common scale, without altering differences in the value's ranges or losing information. The goal is to improve the overall quality of the dataset by re-scaling the dimension of the data and avoiding situations in which some values over-weighting others. Let us say we have age and salary variables that do not have the same scale, and this will cause some issue in machine learning model because most of the ML algorithms are based on Euclidean distance.

$$d(p1, p2) = \sqrt{(x_2 - x_1)^2 + (y_2 - y_1)^2}$$

where $p1$ and $p2$ are two points with (x_1, y_2) and (x_2, y_2) coordinates, respectively. Let us say we have two values.

Age = 30 and 40
Salary = 50,000 and 60,000

One can easily compute and see that salary will be dominated in Euclidean distance and this is not desirable. The solution lies in scaling all the features on a similar scale (0 to 1) or (−1 to 1) There are several ways of scaling the data [7], and few of them are discussed below.

5.6.1 Techniques of Feature Scaling

Machine learning models map from input variables to an output variable. As such, the scale and distribution of the data drawn from domain may be different for each variable. Input variables may have different values (e.g., speed and height in case of flight) that, in turn, may mean the variables have different scales. Differences in the scales across input variables may provide a challenge and difficulty in modeling. An imbalance in associating weightage may result in building an unstable model which would suffer from poor performance during learning. In particular, sensitivity to input values would result in higher generalization error. One of the most common solutions to this problem consists of a simple linear rescaling of the input variables. Many machine learning algorithms perform better when numerical input variables are scaled to a standard range. This includes algorithms that use a weighted sum of the input, like linear regression, and algorithms that use distance measures, like k-nearest neighbors. The two most popular techniques of Feature Scaling are:

1. Standardization
2. Normalization

Both normalization and standardization can be achieved using the scikit-learn library. Let us take a closer look at each in turn.

5.6.1.1 Feature Scaling: Standardization

Standardizing, a dataset involves rescaling the distribution of values so that the mean of observed values is 0 and the standard deviation is 1. This can be thought of as subtracting the mean or centering the data. Standardization assumes that observations fit a Gaussian distribution (bell curve) with a well-behaved mean and standard deviation. It requires that we know or are able to accurately estimate the mean and standard deviation of observable values. One may be able to estimate these values from training data, not the entire dataset. Briefly standardization can be understood as given below.

- Standardization is a popular feature scaling method, which gives data the property of a standard normal distribution (also known as Gaussian distribution).
- All features are standardized on the normal distribution (a mathematical model).

- The mean of each feature is centered at zero, and the feature column has a standard deviation of 1.

To standardize the jth feature, subtract the sample mean μ_j from every sample and divide it by its standard deviation σ_j as given below:

$$x_j^{new} = \frac{x_j - \mu_j}{\sigma_j}$$

scikit-learn implements a class for standardization called scale().

```
# Standardize the data attributes for the Iris dataset.
from sklearn.datasets import load_iris
from sklearn import preprocessing
# load the Iris dataset
iris = load_iris()
print(iris.data.shape)
# separate the data and target attributes
X = iris.data
y = iris.target
# standardize the data attributes
standardized_X = preprocessing.scale(X)
```

Original data

0	1	2	3	
0	5.1	3.5	1.4	0.2
1	4.9	3.0	1.4	0.2
2	4.7	3.2	1.3	0.2
3	4.6	3.1	1.5	0.2

After scaling

	0	1	2	3
0	−0.900681	1.019004	−1.340227	−1.315444
1	−1.143017	−0.131979	−1.340227	−1.315444
2	−1.385353	0.328414	−1.397064	−1.315444
3	−1.506521	0.098217	−1.283389	−1.315444

5.6.1.2   Feature Scaling: Normalization (Min–Max Normalization)

Normalization refers to rescaling the feature between min and max (usually between 0 and 1). To normalize the feature, subtract the min value from each feature instance and divide by the range of the feature (max–min) as shown below.

$$x_j^{new} = \frac{x_j - x_{min}}{x_{max} - x_{min}}$$

where x_j is the original data point, x_j^{new} is the transformed data point, x_{min} is the minimum, and x_{max} is the maximum.

The ML library scikit-learn has a MinMaxScaler class for normalization.

```
from sklearn.preprocessing import MinMaxScaler
X_min_max = pd.DataFrame(MinMaxScaler().fit_transform(X))
X_min_max.head(4)
         0        1        2        3
0.  0.222222 0.625000 0.067797 0.041667
1.  0.166667 0.416667 0.067797 0.041667
2.  0.111111 0.500000 0.050847 0.041667
3.  0.083333 0.458333 0.084746 0.041667.
```

Note: Sometimes machine models are not based on Euclidean distances (ED), we will still need to do features scaling for the algorithm to converge much faster. That will be the case for DT, which are not based on ED. If we skip feature scaling, then the models will run for a longer time.

5.7 Inconsistent Values

In real-world instances, data may contain inconsistent values due to the human error or may be the information was misread while being scanned from a handwritten form or entered by mistake. Due to cut/copy and paste, errors may creep in data. For example, the "Address" field contains the last name. It is therefore always advised to perform data assessment like knowing what the data type of the features should be and whether it is the same for all the data objects.

5.8 Duplicated Values

A dataset may include data objects, which are duplicates of one another. For example, it may happen when the same person submits a form more than once. In this case, we may have repeated information about the person's name or email address. Duplicates may give advantage or bias to the particular data. Therefore, they need to be removed.

5.9   Feature Aggregation

Feature aggregation is performed so as to take aggregated values in order to put the data in a better perspective. Consider the data on electricity usage over a day. Aggregating the usage on an hourly or daily basis will help in reducing the number of observations. This results in saving memory and reducing processing time. Aggregations provide with a high-level view of the data as the behavior of groups or aggregates is more stable than individual data objects.

5.10   Feature Sampling

Sampling is a commonly used technique for selecting a subset of the dataset. In real time, working with the complete data set can turn out to be too expensive considering the memory and time constraints. A sampling technique can be used to reduce the size of the data set that can use a better, but more expensive machine learning algorithm. Sampling should be done in such a manner that the dataset generated after sampling should have approximately the same properties as the original data set, meaning that the sample is representative of the original problem. This involves choosing the correct sample size and sampling strategy. Random sampling dictates that there is nearly equal probability of choosing any particular observation in the dataset. There are two variations of this described below:

5.10.1   Sampling Without Replacement

As each observation is selected, it is removed from the data set of all the objects that from the total dataset.

5.10.2   Sampling with Replacement

Observations are not removed from the original data set after getting selected. This means that observations can get selected more than once. This type of sampling is known as bootstrap sampling.

Although random sampling is a good sampling technique, it can fail to output a representative sample when the data includes object types which are imbalanced. In other words, data includes object types, which vary in ratio. This can cause problems when the sample needs to have a proper representation of all object types, for example, when we have an imbalanced data, i.e., dataset where the number of instances of a class or classes is significantly higher than other class(es).

It is critical that the minority classes are adequately represented in the sample. In these cases, there is another sampling technique, which can be used. It is called stratified sampling. This technique begins with predefined groups of objects. In fact, there are multiple versions of stratified sampling. The simplest version recommends that equal number of observations be drawn from all the groups even though the groups are of different sizes. For more details on sampling techniques, we recommend looking [6].

5.11   Multicollinearity and Its Impact

Multicollinearity occurs in datasets when two or more independent features are strongly correlated with one another. This means that an independent feature can be predicted from another independent variable. For example, body mass index depends on weight and height. Such dependence can impact the ML models adversely. In fact, multicollinearity impacts interpretation capability of the model. Multicollinearity can be a problem because it makes it difficult to distinguish between the individual effects of the independent variables on the dependent variable. The easiest method to identify multicollinearity is to a pair plot or scatter plot and one can observe the relationship between different features. In general, two features within a dataset may be linearly related or may manifest nonlinear relationship. In the case of linear relationship, one can use the correlation matrix also to check how closely related the features are. The closer the value to 1, stronger is the correlation. Another method used to find the multicollinearity is to use variable inflation factors (VIF). VIF determines the strength of the correlation between the independent features. It is predicted by taking a feature and regressing it against every other feature [7]. Although correlation matrix and pair plots can be used to find multicollinearity, their findings only show the bivariate relationship between the independent features. VIF is preferred as it can show the correlation of a variable with a group of other features.

The easiest solution to overcome multicollinearity is to drop one of the correlated features. Dropping features should be an iterative process starting with the feature having the largest VIF value because its trend is highly captured by other features. If we do this, we will notice that VIF values for other features would have reduced too, although to a varying extent.

5.12   Feature Selection

A researcher is often faced with the question: which features one should select to create a predictive model? This is a difficult question. It requires deeper understanding or knowledge of the problem domain. Sometimes, it is possible to automatically select those features in the data that are most useful or most relevant for the problem we are solving. This is a process known as feature selection.

Feature selection and feature engineering are two very important aspects for the success of machine learning models. So far, we have discussed data preprocessing or feature engineering techniques and now we discuss feature selection.

Sometimes, it is observed that datasets are small, but more often, they are extremely large in size. Then, it becomes challenging to process the datasets to avoid processing bottleneck. So, what makes these datasets this large? Well, it is the features or the dimensions inherent in the data. With a greater number of features, the larger are the dataset's dimension. For example, in the case of text mining, there could be millions of features with dataset size of the order of several GB. One might wonder with a commodity computer in hand how to process these types of datasets.

Often, in high-dimensional datasets, there are several irrelevant and unimportant features. The contribution of these types of features is often less toward predictive modeling in comparison to the critical features. The unimportant features may have no contribution or zero contribution. These features can cause a number of problems in predictive models. Some of these problems are listed below:

- Unnecessary storage allocation
- Act as a noise for which the ML model can perform poorly
- Long training time

So, what is the solution? The best solution is feature selection. Feature selection is the process of selecting most significant/important feature from a given dataset. The subset selection belongs to the class NP-hard. The techniques can be mainly categorized into two branches, greedy algorithms and convex relaxation methods [8].

5.12.1   Importance of Feature Selection

Machine learning models work on a simple rule—garbage in, garbage out. By garbage here means noise in data. As discussed above, the high-dimensional dataset can lead to lot of problems, namely long training time, model could be complex which in turn may lead to over-fitting. We do not need to use every feature that is present in the dataset. One can assist modeling by feeding in only those features that are really important. Sometimes, less is better. In a case with very high dimensions, there are several redundant features that do not contribute much but are simply extensions of the other important features. These redundant features do not contribute to the model's predictive capability. Clearly, there is a need to remove these redundant features from the dataset to get the most effective predictive modeling performance. We can summarize the objectives of feature selection as follows:

- It enables faster training of a model.
- It reduces the complexity of the model, and it becomes easier to interpret and avoids over fitting.
- It improves the prediction accuracy of a model.
- Less data to store and process.

Here it would be worthwhile to mention Prof. Pedro Domingos (University of Washington) quotation:

"At the end of the day, some machine learning projects succeed and some fail. What makes the difference? Easily the most important factor is the features used."

It is important to understand the difference between dimensionality reduction and feature selection. Quite often, feature selection is mistaken with dimension reduction. But they are different. Both methods tend to reduce the number of features in the dataset, but a dimensionality reduction method does so by creating new combinations of attributes, whereas feature selection methods include and exclude features without changing them. Some examples of dimensionality reduction methods are principal component analysis (PCA), linear discriminant analysis (LDA), etc. We will discuss these techniques later in the chapter. In this section, we are concentrating on feature selection.

5.12.2   How Many Features to Have in the Model?

One important point to consider is the tradeoff between predictive accuracy vs. model Interpretability. This is because when we use large number of features, the predictive accuracy may go up while model interpretability goes down. On the other hand, if we have a smaller number of features, then it is easier to interpret the model. We are less likely to over-fit. However, it may give relatively lower predictive accuracy.

5.12.3   Types of Feature Selection

There are various methodologies and techniques that can be used to generate a subset from a given dataset. Figure 5.2 shows the feature selection methods used in ML. Next, we will discuss these methods.

5.12.3.1   Filter Method

Figure 5.3 describes feature selection method. Filter methods are generally used as a preprocessing step. Using this method, the predictive power of each individual feature is evaluated. The selection of features is independent of any ML algorithm.

Fig. 5.2 Various feature
selection methods

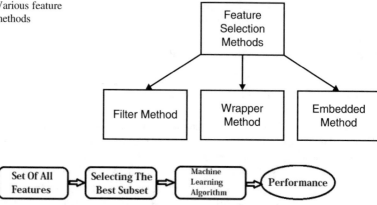

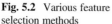

Fig. 5.3 Filter-based feature selection method

Table 5.1 Various correlations between different variable types

Feature\output	Continuous	Categorical
Continuous	Pearson's correlation	Linear discriminant analysis (LDA)
Categorical	ANOVA	Chi-square

Filter method uses assessment criterion, which includes distance, information, dependency, and consistency. The filter method uses the principal criteria of ranking technique and uses rank ordering method for feature selection. The reason for using ranking-based method is its simplicity and produce relevant features. This method will select relevant feature before feeding them into the ML algorithm. Features give rank on the basis of statistical scores that tend to determine the feature's correlation with the target feature. The correlation is a subjective term. The features with the highest correlation are the best.

For example: Y is target variable and $(X1, X2, X3, \ldots Xn)$ are independent variables. We find out the correlation between target variable with respect to independent variables. $(Y \rightarrow X1)$, $(Y \rightarrow X2)$, $(Y \rightarrow X3), \ldots (Y \rightarrow Xn)$. So, the features, which have highest correlation with Y, will be selected as best features.

Finding correlation coefficient depends upon the type of variable, and readers are referring to Table 5.1.

- Pearson's coefficient: It is used to measure linear dependency between two continuous variables. Its value varies from -1 to $+1$. The closer the value to 1, the stronger is the correlation. Sign indicates the direction of the relation. There are other correlations defined in the literature {Ref}.
- LDA: It is used to find the linear combination of features that characterizes or separates two or more classes of a categorical variable.
- ANOVA: ANOVA stands for analysis of variance. It is similar to LDA except for the fact that it is operated using one or more categorical independent features and

one continuous dependent feature. It provides a statistical test of whether the means of several groups are equal or not.
- Chi-square: It is a statistical test applied to the groups of categorical features to evaluate the likelihood of correlation of association between them.

5.12.3.2   Wrapper Methods

Wrapper methods use combinations of features to determine predictive power. In wrapper methods shown in Fig. 5.4, a subset of features is used along with a potential model. Based on the inferences drawn from the model, a particular feature is added or removed from the model. The wrapper method will often find the best combination of features. The problem with these methods though is that they are computationally expensive. It is a NP complete problem. It is not recommended that this method be used on a large number of features. Common wrapper methods include: Subset selection, Forward stepwise, and Backward stepwise. We will discuss them below.

- *Subset selection*: In subset selection, the model is fitted with each possible combination of N features, and the best model is selected. Let us say we have N number of independent features in a dataset, so total number of models in the subset selection will be 2^N models. Subset selection requires massive computational power to execute.
- *Forward selection*: Forward selection is an iterative method in which model is started having no feature, i.e., it starts with no variable in the model. In each iteration, we keep adding the feature that improves the model till an addition of a new variable does not improve the performance of the model. In this method, once the feature is selected, it never drops in second step. Choose the model among the bests of model based on residual sum of squares (RSS) or adjusted R square. In forward selection, selection is the constrained as a predictor that is in model never drops. So, selection models in forward selection becomes $1 + N$ $(N + 1)/2$ which is a polynomial complexity. In this case, computational power is reduced substantially as compared to subset selection (Fig. 5.5).
- *Backward selection*: It works in the opposite direction in that it eliminates features. Because they are not run on every combination of feature, they are orders of magnitude less computationally intensive than straight subset selection and is similar to forward selection. In this method, all the features are considered

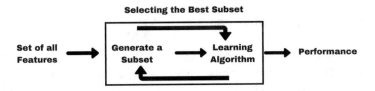

Fig. 5.4 Wrapper Methods

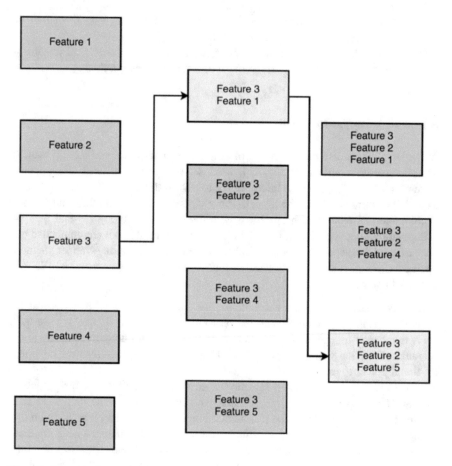

Fig. 5.5 Forward selection method

at the start and remove the least significant feature at each iteration that improves the performance of the model. The process is repeated until no improvement is observed on the removal of features (Fig. 5.6).

Recursive feature elimination: It is a greedy optimization algorithm that aims to find the best performing feature subset. It repeatedly creates models and keeps aside the best or the worst performing feature at each iteration [9].

5.12.3.3   Embedded Methods (Fig. 5.7)

Embedded method is implemented by algorithms that have their own built-in feature selection methods. The most common types of embedded feature selection methods are known as regularization methods or shrinkage methods. Regularization has inbuilt penalty functions to penalize and identify features which are not important.

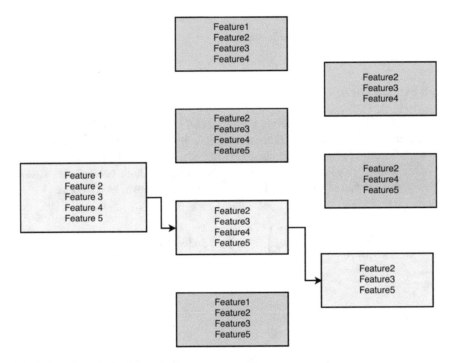

Fig. 5.6 Backward selection method

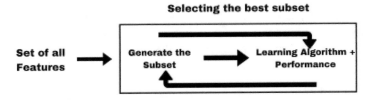

Fig. 5.7 Embedded method

Regularization methods introduce additional constraints into the optimization of a predictive model that biases the model toward lower complexity. This controls the value of parameters. In other words, basically not so important features are given very low weight. This technique discourages learning a more complex or flexible model, so as to avoid the risk of over-fitting when one attempts prediction on the new data. Some of the most popular examples of these methods are LASSO, Elastic net, and RIDGE regression. Ridge and Lasso regression are powerful techniques generally used for creating parsimonious models in the presence of a 'large number of features. Though Ridge and Lasso might appear to work toward a common goal, the inherent properties and practical use cases differ substantially. These methods work by penalizing the magnitude of coefficients of features along with minimizing the error between the predicted and actual observations. The key difference is in how they assign penalty to the coefficients.

LASSO Regression

- LASSO stands for least absolute shrinkage and selection operator. In this method, few of the coefficients of predictors shrink to zero, that is why we drop or reject such features. In this method, the following function is minimized. This variation differs from ridge regression only in penalizing the high coefficients. It uses modulus instead of squares of β, as its penalty. It is known as L1 norm.

$$\sum_{i=1}^{n}\left(y_i - \sum_{j=1}^{p}\beta_j x_{ij}\right)^2 + \lambda\sum_{j=1}^{p}|\beta_j| = RSS + \lambda\sum_{j=1}^{p}|\beta_j|$$

Ridge Regression

- RSS is modified by adding the shrinkage quantity as shown the equation. This adds a penalty, which equals the square of the magnitude of coefficients. All coefficients are shrunk by the same factor (so none of the features are eliminated). Ridge regression is very similar to least squares, except that the coefficients are estimated by minimizing a slightly different quantity. In particular, the ridge regression coefficient estimates betas are the values that minimizes.

$$\sum_{i=1}^{n}\left(y_i - \beta_0 - \sum_{j=1}^{p}\beta_j x_{ij}\right)^2 + \lambda\sum_{j=1}^{p}\beta_j^2 = RSS + \lambda\sum_{j=1}^{p}\beta_j^2$$

where $\lambda \geq 0$ is a tuning parameter, to be determined separately. The tuning parameter (λ) controls the strength of the penalty term and decides how much to penalize the flexibility of the model. When $\lambda = 0$, ridge regression equals least squares regression. If $\lambda = \infty$, all coefficients shrunk to zero. The ideal penalty is therefore somewhere is between 0 and ∞ and selecting a good value of λ is critical. Cross validation comes in handy for this purpose. The coefficient estimates produced by this method are also known as L2 norm.

For further reading, readers are referred to [10, 11].

5.13   Dimensionality Reduction

Usually, real-world data have a large number of features. For example, image-processing problem may have thousands of features. Features are also referred as dimensions. Dimensionality reduction aims at reducing the number of features for processing. This is called feature subset selection or simply feature selection.

Conceptually, dimensions refer to the number of geometric planes on which the dataset lies. This number could be too large sometimes for a realistic visualization. Larger the number of such planes, greater is the complexity of the dataset. Data analysis task becomes significantly harder as the dimensionality of the data increases. As the number of dimensions increases, the number of planes occupied by the data increases. It should be noted that higher the dimensionality of data, greater is the sparsity. This leads to difficulty in modeling and visualizations.

Dimension reduction maps the dataset to a lower dimensional space. The basic objective of using dimension reduction techniques is to create new features that are combination of the original features. In other words, the higher-dimensional feature space is mapped to a lower-dimensional feature space. Techniques like forward feature selection, backward feature selection models like Random Forest can also be used for dimension reduction discussed earlier. Here we will briefly touch upon principal component analysis (PCA), singular value decomposition (SVD), T-SNE, and discriminant analysis techniques to achieve dimensionality reduction. These methods fall under feature extraction methods. Using these methods, we extract or engineer new features from the original features in the given dataset. Thus, the reduced subset of features will contain newly generated features that were not part of the original feature set.

5.13.1   Principal Component Analysis (PCA)

PCA was introduced by Pearson [12]. PCA is a mathematical procedure that transforms a number of correlated features into a (possibly smaller) number of uncorrelated features which are considered to be principal components. PCA, even though invented more than a century ago, has proven itself to be one of the most important and widely used algorithms in modern data science. It has been gainfully used for visualization of high-dimensional data, unsupervised learning, and dimensionality reduction. Its broad appeal has meant that it has become a mainstay in numerical computing and AI software libraries alike.

PCA is a method that rotates given dataset in a way such that the rotated features are statistically uncorrelated. This rotation is often followed by selecting only a subset of the new features, depending upon how important they are for explaining the data. It primarily looks at the correlations within the data. This technique is particularly useful in processing data where multi-collinearity exists between features or when the dimensions of features are high. In other words, PCA works on the

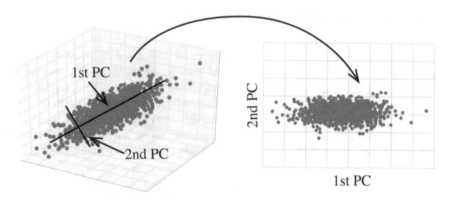

Fig. 5.8 Transformation of data with principal component

premise that while the data is in a higher-dimensional space, it may be possible to map it into a data representation in a lower dimension space such that the variance of the data in the lower dimensional space is minimum. While variance measures a random variable's spread, whereas the co-variance measures the extent/spread of one random variable with respect to another random variable. All that PCA tries to do is to replace correlated dimensions and less variance features with their linear combinations. The mathematical objective of PCA is to retain those dimensions that proffer maximum variance (important features). It gives us a new set of dimensions that are orthogonal and are ranked in the order of higher variance. Resulting dimensions having high variance are called principal components. It should be mentioned here that mean normalization and feature scaling are a must before performing the PCA. This ensures that the variance of a component is not affected by the disparity in the range of values. PCA is different from linear regression. In linear regression, the goal is to predict a dependent variable, given independent variable, and we minimize the prediction error. PCA does have a response variable, and it ensures a feature reduction by minimizing the projection error (Fig. 5.8).

PCA is normally implemented in one of two ways:

1. Using singular value decomposition
2. Through eigenvalue decomposition of the covariance matrix

Steps involved:

1. Perform standardization on data (X). Calculate the co-variance matrix of data.
2. Calculate the eigen-values and eigen-vectors over co-variance matrix.
3. Choose the principal components (PC).
4. Construct new feature data set from chosen components:

Final $\rightarrow$ Transpose (PC). Transpose (X).

Limitations:

- It finds linear relationships. Sometimes, it needs kernels that are nonlinear.
- It gives orthogonal vectors. Sometimes, we might have data having variance at two different directions but not orthogonal. In that case, PCA just gives orthogonal vectors.
- It may not preserve the shape of the data.

Suggestions for using PCA:

- Speed up e-learning algorithm by reducing the number of features by applying PCA and choosing top-k principal components to maintain 99% variance. PCA should be applied on the training data only.
- If using PCA for visualization, it does not make sense to choose $k > 3$.
- Usage of PCA to reduce over-fitting is not advised. The reason it works well in some cases is because it reduces the number of features leading to reduction in the variance and enhances the bias. But often there are better ways of doing this by using regularization and other similar techniques than use PCA. This would be bad application of PCA.
- PCA can also be used in cases when the original data is too big for disk space. In such cases, compressed data will give some benefits of space saving by dimensionality reduction.

Code for PCA:
Scikit-learn provides a good module for PCA. Here n components are the number of dimensions we want to reduce to. Explained_variance_ratio property gives quantity of variance each feature gives. X is the input dataset.

```
from sklearn.decomposition import PCA
pca = PCA(n_components=4)
pca.fit(X)
print(pca.explained_variance_ratio_)
```

5.13.2   *Linear Discriminant Analysis*

LDA is most commonly used as dimensionality reduction technique in the preprocessing step in machine learning applications. It was originally developed in 1936 by Fisher [13]. The goal is to project a dataset onto a lower-dimensional space with good class separability in order to avoid over-fitting and also reduce computational costs. LDA is a supervised learning technique for dimension reduction and aims to maximize the distance between the mean of each class and minimize the spread within the class itself. This is a good choice because maximizing the distance between the means of each class when projecting the data in a lower-dimensional space can lead to better classification results. It is assumed that input data follows a Gaussian distribution.

Code for LDA:

```
from sklearn.discriminant_analysis import LinearDiscriminantAnalysis as LDA
lda = LDA(n_components=4)
X_lda = lda.fit_transform(X)
```

5.13.3   t-Distributed Stochastic Neighbor Embedding (t-SNE)

t-SNE is a technique for dimensionality reduction that is particularly well suited for the visualization of high-dimensional datasets. It was developed by Maaten and Hinton [14–16]. It is a nonlinear dimensionality reduction technique well-suited for embedding high-dimensional data for visualization in a low-dimensional space of two or three dimensions. The t-SNE algorithm comprises two main stages.

- t-SNE constructs a probability distribution over pairs of high-dimensional objects in such a way that similar objects are assigned a higher probability while dissimilar points are assigned a relatively much lower probability.
- t-SNE defines a similar probability distribution over the points in the low-dimensional map, and it minimizes the Kullback–Leibler divergence (KL divergence) between the two distributions with respect to the locations of the points in the map. The original algorithm uses the Euclidean distance between objects as the base of its similarity metric. t-SNE has been used in a wide range of applications, including computer security, music analysis, cancer research, and biomedical signal processing. Those who are interested in knowing the detailed working of an algorithm can refer to [16].

The details of these techniques are beyond the scope of this book. Readers are referred to [17–19].

5.14   Dealing with Imbalanced Data

Imbalance in dataset can introduce unique challenges to the learning algorithm. The learning challenge manifests as a form of class imbalance. The name speaks for itself; imbalanced data typically refers to the datasets where the number of observations per class is not equally distributed. Often, we will have a large amount of observations for one class (referred to as the majority class), and much fewer observations for one or more other classes (referred to as minority classes). One can have a class imbalance problem on two-class classification problems as well as multi-class classification problems. Imbalanced data is not always a bad thing, and in practice, there is always some degree of imbalance. That said, there will be minimal impact on the model performance if the level of imbalance is relatively low.

There are problems where a class imbalance is not just common: it is bound to happen. For example, in areas such as credit card transaction: a fraudulent or authentic—in this case there may be thousands of authentic transactions for every fraudulent transaction, that is quite an imbalance and may be a concern. In imbalanced dataset, machine learning models tend to have frequency bias in which they place more emphasis on learning from data observations which occur more commonly. Therefore, it is imperative to choose the evaluation metric of learning model correctly. If it is not done, then one might end up adjusting/optimizing a useless parameter. In business, this may lead to a complete waste. The evaluation of ML algorithms may show why a particular ML algorithm does not perform well with imbalanced data. It is the case where accuracy measures tell the story that one has excellent accuracy (such as 90%), but the accuracy is only reflecting the underlying class distribution [20].

Although machine learning algorithms have shown great success in many real-world applications, the problem of learning from imbalanced data is yet to become the state-of-the-art. There are three main problems imposed by imbalanced data [21]. They are as follows:

1. *The machine problem*: ML algorithms are built to minimize errors. Since the probability of instances belonging to the majority class is significantly high in imbalanced dataset, the algorithms are much more likely to classify new observations to the majority class.
2. *The intrinsic problem*: In real life, the cost of false negative is usually much larger than false positive, yet ML algorithms penalize both with similar weightage.
3. *The human problem*: This is in the context of banking operation. In credit risk, common practices are often established by experts, rather than empirical studies [22]. This is surely not optimal, given that population might be very different from the other bank's population. Therefore, what works in a certain loan portfolio might not work in others.

There are several articles addressing the issue with imbalanced data. We will discuss few of the solutions.

5.14.1   Use the Right Evaluation Metrics

Applying inappropriate evaluation metrics for model generated using imbalanced data can be dangerous. Accuracy is not a good measure in this case as it will classify majority of the class and accuracy will be high. In this case, other alternative evaluation metrics may be applied such as:

- Precision/specificity
- Recall/sensitivity

- F1 score
- AUC; area under ROC
- Mathew's correlation coefficient

These have been discussed in Chap. 2 of the book.

5.14.2   Sampling-Based Approaches

This can be roughly classified into three categories:

1. Oversampling: by adding more of the minority class so it has more effect on the machine learning algorithm.
2. Under sampling: by removing some of the majority class so it has less effect on the machine leaning algorithm.
3. Hybrid: a mix of oversampling and under sampling.

In 2002, sampling-based algorithm called SMOTE (Synthetic Minority Over-Sampling Technique) was introduced that tries to address the class imbalance problem. It is one of the most adopted approaches due to its simplicity and effectiveness. It is a combination of oversampling and under sampling, but the oversampling approach is not by replicating minority class but constructing new minority class data instance via an algorithm.

5.14.3   Algorithm Based Approach

As mentioned above, ML algorithms penalize FP and FN equally. A way to counter that is to modify the algorithm itself to boost predictive performance on minority class. This can be executed through either recognition-based learning or cost-sensitive learning [23, 24].

The class imbalance problem is a common problem affecting ML models due to having disproportionate number of class instances in practice. The detail of this is beyond the scope of this book. Interested readers may look into literature [25–29].

5.15   Points to Ponder

1. What is the need for data clean-up before feeding it to ML algorithms?
2. Why is feature scaling necessary?
3. What is the impact of multicollinearity?

References

1. https://towardsdatascience.com/data-preprocessing-concepts-fa946d11c825.
2. Browniee, J. (2020). *Data preparation for machine learning*. Vermont: Machine Learning Mastery.
3. https://machinelearningmastery.com/data-preparation-for-machine-learning/.
4. https://machinelearningmastery.com/framework-for-data-preparation-for-machine-learning/.
5. Senawi, A., Wei, H. L., & Billings, S. A. (2017). A new maximum relevance-minimum multicollinearity (MR-mMC) method for feature selection and ranking. *Pattern Recognition, 61, 67*, 47.
6. https://medium.com/@analytics.
7. https://www.analyticsvidhya.com/blog/2020/03/what-is-multicollinearity/.
8. Garey, M. R., & Johnson, D. S. (1979). *Computers and intractability. A guide to the theory of NP-completeness* (A series of books in the mathematical sciences). San Francisco, CA: W. H. Freeman and Co..
9. https://www.analyticsvidhya.com/blog/2016/12/introduction-to-feature-selection-methods-with-an-example-or-how-to-select-the-right-variables/.
10. Hastie, T., Tibshirani, R., & Friedman, J. (2009). Elements of Statistical Learning (Data Mining, Inference and Prediction), Springer.
11. https://www.analyticsvidhya.com/blog/2016/01/ridge-lasso-regression-python-complete-tutorial/.
12. Pearson, K. (1901). On lines and planes of closest fit to systems of points in space. *Philosophical Magazine*, Series 6, *2*(11), 559–572.
13. Fisher, R. A. (1936). The use of multiple measurements in taxonomic problems. *Annals of Eugenics, 7*(2), 179–188.
14. van der Maaten, L. J. P. (2014). Accelerating t-SNE using tree-based algorithms. *Journal of Machine Learning Research, 15*(October), 3221–3245.
15. van der Maaten, L. J. P., & Hinton, G. E. (2012). Visualizing non-metric similarities in multiple maps. *Machine Learning, 87*(1), 33–55.
16. van der Maaten, L. J. P., & Hinton, G. E. (2008). Visualizing high-dimensional data using t-SNE. *Journal of Machine Learning Research, 9*(November), 2579–2605.
17. https://machinelearningmedium.com/2018/04/22/principal-component-analysis/.
18. Raschka, S. (2014). Introduction to linear discriminant analysis. https://sebastianraschka.com/Articles/2014_python_lda.html.
19. In depth: Principal component analysis. https://jakevdp.github.io/PythonDataScienceHandbook/05.09-principal-component-analysis.html.
20. https://machinelearningmastery.com/tactics-to-combat-imbalanced-classes-in-your-machine-learning-dataset/.
21. https://medium.com/james-blogs/handling-imbalanced-data-in-classification-problems-7de598c1059f.
22. Crone, S., & Finlay, S. (2012). Instance sampling in credit scoring: An empirical study of sample size and balancing. *International Journal of Forecasting, 28*(1), 224–238.
23. Drummond, C., & Holte, R. C. (2003). Cost-sensitive classifier evaluation using cost curves. In T. Washio, E. Suzuki, K. M. Ting, & A. Inokuchi (Eds.), *Advances in knowledge discovery and data mining. PAKDD 2008* (Lecture notes in computer science) (Vol. 5012). Berlin, Heidelberg: Springer.
24. Elkan, C. (2001) The foundations of cost-sensitive learning. In: *Proceedings of the seventeenth international joint conference on artificial intelligence (IJCAI'01)*.
25. http://www.chioka.in/class-imbalance-problem/.

26. More, A. (2016). Survey of resampling techniques for improving classification performance in unbalanced datasets. arXiv:1608.06048.
27. https://medium.com/james-blogs/handling-imbalanced-data-in-classification-problems-7de598c1059f.
28. https://www.kdnuggets.com/2017/06/7-techniques-handle-imbalanced-data.html.
29. Kuhn, M. (2018). *Applied predictive modeling*. New York: Springer.

Chapter 6
Information Security and Cloud Computing

6.1 Information Security Background and Context

Cloud computing exacerbates computer security issues arising from multi-tenancy and open access due to multiple users sharing resources. We will start by looking at traditional information security challenges and then see how these evolve into threats in the cloud [1]. Information security includes the following three fundamental activities:

1. *Access control*: Access control is usually addressed at the operating system level with a login step. This includes both the initial entrance by a participant and the reentry of that participant or the access of additional participants. Note that a participant can be an individual or some computer process. The first function encountered is access control, i.e., who can rightfully access a computer system or data. The access control can be resolved at a hardware level with a special access device such as a dongle connected to the USB port or built in security keys. An example of access control at the application level is the setting of cookies by a browser.

2. *Secure communications*: Secure communication includes any transfer of information among any of the cloud participants. Most commonly recognized function of a secure system is the encryption algorithm. One of the challenges in a secure system is the management of encryption keys. At the hardware level, the communication encryption device can be implemented at the I/O port. At the operating system level, encrypted communication can be implemented in secure driver software. At the application level, the encryption algorithm is implemented in routines performing secure communication. Some of the other functions and issues for security systems are:

   (a) Hashing (for checking data integrity)
   (b) Identity authentication (for allowing access)
   (c) Electronic signatures (for preventing revocation of legitimate transactions)

© The Author(s), under exclusive license to Springer Nature Switzerland AG 2021   143
P. Gupta, N. K. Sehgal, *Introduction to Machine Learning in the Cloud with Python*,
https://doi.org/10.1007/978-3-030-71270-9_6

(d) Information labeling (for tracing location and times for transactions)
(e) Monitors (for identifying potential attacks on the system)

Each of these functions affects the overall security and performance of the system. The weakest security element for any function at any level delimits the overall security. The weakest function or element is not only determined by technical issues (such as length of passwords) but also by user acceptance.

3. *Protection of private data:* Protection of private data includes storage devices, processing units, and even cache memory. In addition to accepting these security processes, a user may have privacy concerns. Specifically, when more information is required to ensure proper access, the more private information is available to the security checking system. The level of security required is generally at four levels, as enumerated below:

(a) Ease of access is more important for low security activities, such as reading advertisements.
(b) More difficult access is required for medium security such as bank accounts.
(c) High security is required for high-value corporate proprietary computations, such as design data for next-generation product.
(d) Very strict and cumbersome access procedures are expected for nuclear weapons applications.

These examples provide a clue to security in a cloud computing environment with shared resources. Specifically, in the shared computing environment, applications run at a variety of security levels. Security solutions must also consider the tradeoffs of security versus performance. Some straightforward increases in the security cause inordinate degradation of performance, due to computations required for real-time decryption, etc.

In a typical public cloud interaction, once a user access is granted, secure communication is the next function, which requires encryption. As described previously, the security implementations can be done at multiple levels for each of the functions. Because security is a multi-function multilevel problem, high-level security operations need access to low-level security measurements. This is true for measuring both performance and security in a cloud. Following changing environmental factors directly affect the evolution of information security. One factor has been, and continues to be, the computer power available to both sides of the information security battle. Computing power continues to follow Moore's law with increasing capacity and speeds increasing exponentially with time. Therefore, while the breaking of a security information security and cloud computing system with brute force may take many years with the present computer technology. In only a few years, the computer capacity may be available to achieve the same break-in within real time. Another environmental factor is the increasing number of people needing information security. The world has changed from a relatively modest number of financial, governmental, business, and medical institutions having to secure information to nearly every business and modern human needing a support for information security. The sheer number of people needing information security at different levels has increased the importance of security. The third environmental change that has a significant impact on the security is the sharing of information resources. That is the crux of this chapter. Specifically, we will describe the information security challenges caused by the spread of data centers and cloud computing. More people across the world are accessing Internet not just through PCs and browsers, but using cell phones, IoT devices, and mobile applications. This has dramatically increased the risks and scale of potential damage caused by realization of a security threat on the Internet.

6.2 Privacy Issues

Even before the advent of public cloud, computing was mainly done behind the enterprise firewalls. Most threats in this scenario were from the insider attackers. There was a need to segregate data on a need to know bases. Different departments, such as accounting or engineering, had their own different disk partitions that were accessible only to the employees of the respective departments. Here the system administrator was considered to be in the Trusted Compute Boundary (TCB), which meant that he or she had access to all the data. Any personnel data is partitioned with public and private visibility, such that former type information was freely available to all employees of the enterprise. This included names of employees, their job title, office location, etc. Other private information such as their date of birth, social security numbers, and salary information was considered secret, thus accessible only to the human resources department or the concerned employee. Any information leaks were handled with strict disciplinary actions, including up to termination. However, the extent of leaks was limited to the employees.

With the adoption of public cloud, where data from many entities is aggregated and resides in remote servers, the potential damage from leaks is enormous. An example is the Equifax hack [2], where the personal information of 147 million consumers was accessed. It happened in the data centers of a credit reporting agency that assesses the financial health of nearly everyone in the United States. The Equifax breach investigation highlighted a number of security lapses that allowed attackers to enter supposedly secure systems and exfiltrate terabytes of data. To understand how this happened, let us look at the following sequence of events:

- The company was initially hacked via a consumer complaint web portal, with the attackers using a widely known vulnerability that should have been patched, but due to failures in Equifax's internal processes wasn't.
- The attackers were able to move from the web portal to other servers because the systems were not adequately segmented from one another, and they were able to find usernames and passwords stored in plain text that then allowed them to access additional systems.

- The attackers pulled data out of the network in encrypted form undetected for months because Equifax had crucially failed to renew an encryption certificate on one of their internal security tools.
- Equifax did not publicize the breach until more than a month after they discovered it had happened; stock sales by top executives around this time gave rise to accusations of insider trading.

Thus, it is important for any private information, or personally identifiable information (PII), to be not segmented and encrypted as was done by Equifax, but also the keys or certificates be properly maintained in a secure manner.

6.3 Security Concerns of Cloud Operating Models

Traditional computing environments had a clear distinction between "inside" and "outside." If we consider a typical computer user such as Alice, then an "inside" might be in Alice's office or "inside" the bank building. With the dawn of networks, and especially the Internet, the networks were partitioned as "inside the firewall" and "outside" which could be anywhere. This is one of the differences between a public cloud and a private cloud. Secure communication was only needed when the communication was from "inside" to "outside." With cloud computing, "inside" is not clearly defined as computers in data centers across different geographies are pooled together to appear as a large virtual pool. This is true for both public and private clouds because a private cloud can use a Virtual Private Network (VPN) that uses the open Internet to connect services only to a restricted category of internal clients. So, depicted in Fig. 6.1, an attacker such as Eve and Malory could have access to the same set of physical resources that Alice is using in a cloud. Some people more easily trust a private cloud. Remember that within a private cloud,

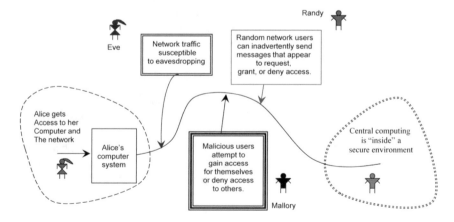

Fig. 6.1 Various cloud players and security attack scenarios

eavesdropping Eve and malicious Malory could be a coworker or someone external. Also, in a cloud environment, the unauthorized access could be by some random Randy that inadvertently slipped through the access barrier. The monitoring and response for security purposes must not only consider the level of secrecy and impact of a breach but also the category of intruders.

As is shown in Fig. 6.1, the definition of "inside" the security perimeter and "outside" the perimeter is clear. When Alice wishes to get information to Bob, she can just walk over and stay inside the secure perimeter. Eavesdropping by Eve requires a much greater sophistication than the effort to secure the communication. For example, Alice may post documents on her bulletin board facing a clear window open to Eve's apartment across the street, providing her a way to view remotely. Even in the traditional environment information security leaks do occur. Phone conversations, documents, and removable storage media did provide opportunities for information from within the secure environment to get to unauthorized individuals. However, usually enforcing policies based upon the physical boundary was sufficient to provide information security.

The Internet created a new issue—that is connecting secure islands of information to each other via a very open channel. The computing center with large computing and storage resources still has controlled access and activities. This creates a large island of security where the major resources can still be controlled and monitored with humans and devices. Although the system operators have access via uncontrolled hardware lines, identical to regular user access lines. Unlike the traditional case, there is no casual monitoring by coworkers. As has been said in the cartoon world, "On the Internet nobody knows your real identity." Also, the Internet provides an intruder with unlimited attempts to gain access. After an intruder has failed to gain access, the failed attempt cannot be attached to the intruder. Each attempt may appear as a first attempt as IP addresses can be spoofed. Time to attempt repeated access is greatly reduced. Procedures to stop these intruders also impact legitimate users. For example, one can stop the high number of repeated access attacks by limiting the number of false attempts by locking an account after some number of false attempts. However, this can prevent a legitimate user from accessing her account. This can lead to another form of security attack called denial of service. The idea here is not to gain access, but to prevent legitimate users' access by hitting the access attempt limit and thus locking out (or denying service) to legitimate users [3].

6.4 Secure Transmissions, Storage, and Computation

Cloud computing introduces the potential for extra access to resources by unauthorized parties. This is shown in Fig. 6.2. The communication between islands of security (shown in green) through a sea of openness (in light blue background) is solved by encrypting all of the data that traverses across the open sea. A particularly extreme example is attacking main memory persistence. A solution to this problem

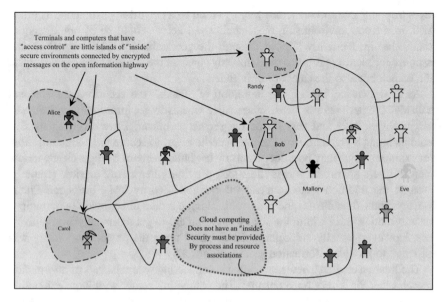

Fig. 6.2 Information security and cloud computing on the Internet

has been addressed for the traditional and Internet cases [4]. Specifically, research on cryptographic methods for authenticating data stored in servers [5]. In cloud computing, we can have a process swapping between users that complicates the direct assignment of main memory to a particular process. Hence, on the fly encryption of the data bus is required between the on-chip cache and the main memory. Additionally, the encrypted virtual storage of the main memory could lose data on restart. For example, a very long and complex simulation may be check pointed, but a power glitch requires a restart. One issue for the customer or end user is to check whether the data in the cloud is available and correct when requested.

Cloud providers are essentially semi-trusted suppliers. Without undue burden on the system, the data availability and integrity must be checked [6]. This is demonstrated via Proof of Retrievability (POR) checks [7]. The data and execution of VMs must be monitored [8]. Denial of service attacks can occur in clouds implemented with multi-core servers by targeting cache accesses [9].

6.5 A Few Key Challenges Related to Cloud Computing and Virtualization

In a traditional computing center, anything inside the physical boundaries is considered secure. However, a cloud does not have clear boundaries because of its distributed nature. In cloud computing, users are concerned about the privacy of their data because they do not know where their data is being processed. A key

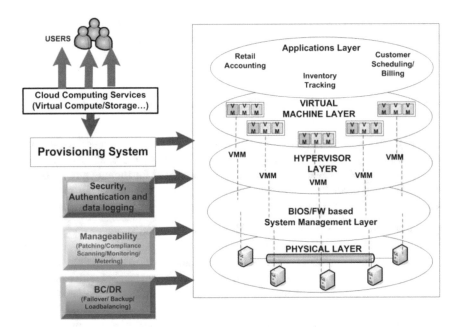

Fig. 6.3 Security inside a data center

problem that customers face is the trust and confidentiality of their data from the end-user devices being projected into a cloud supporting data center. In a computing system inside a datacenter, as shown in Fig. 6.3, components are connected by various buses. An adversary could physically tap those busses to gain information. This is possible as an attacker could replace some modules or insert some components in a computer, for example, during the delivery phase. Hence, those busses are not trusted. It is hard for each and every administrator or privileged personnel from a cloud service provider to get a universal security clearance, especially when a third-party service provider is involved. The providers can use tools such as "QUIRC: A Quantitative Impact and Risk Assessment Framework for Cloud Security" to evaluate the security of their enterprise [10]. It is important for cloud service providers to make their customers more confident in their claim of data privacy and integrity.

Providers using cloud may face additional challenges due to dynamic provisioning of multiple virtual servers on physical machines to achieve the economy of scale. This implies that data from potentially competing sources could reside on the same disk or memory structure, and through an accident or by design, a computer process can violate the virtual boundary to access a competitor's data. Furthermore, a write or data-trashing activity may occur that can go un-noticed. The damage of this can be contained by having secure backups, authentication and data logging for each user's activity in a cloud data center. Another specific problem of multiple virtual machines (VMs) is information leaks. These leaks have been used to extract RSA and AES encryption keys [11, 12]. However, it may create another problem of a large volume of logs, storage, privacy issues of who else can see a user's accesses.

6.6   Security Practices for Cloud Computing

Securing the cloud is akin to the problem of boiling the ocean, i.e., it will require a lot of time and energy, even then a single breach can leak a lot of data as was seen in the case of Equifax [2]. However, cloud security is essential for running any AI or ML workloads to protect the intellectual property (IP). Below are seven of the best security practices to keep cloud environment safe [13]:

1. *Shared Cloud Security Responsibilities*: Both the cloud service provider (CSP) and user are responsible for security in the cloud. A service level agreement (SLA) specifies the aspects of cloud security that the user is responsible for, and it also spells out obligations of the cloud vendor. It should be read with care and, if necessary, negotiated to change the terms to ensure a mutually acceptable level of security.
2. *Data encryption in the cloud*: It is a customer's responsibility to ensure that that data transmission and storage in the cloud is encrypted. A cloud environment needs to provide tools for data encryption at rest and during transmission. This should be checked before migrating workloads to a public cloud.
3. *Establishing cloud data deletion policies*: These days many businesses are adopting multi-cloud strategies [14]. This often requires migration to a new cloud or back to an on-premise server. Hence, there will be data in the public cloud environment that is no longer needed and should be deleted in a safe manner. This requires all storage and memory content to be zeroed out, not just freeing up the space for next computing process. This is recommended because a hacker can use undelete function to read the previous memory or storage content. This also includes backups that the cloud vendor may have taken.
4. *Managing access control*: Different types of users should have different rights and specific access policies in the cloud. This includes vendor's system administrators who may not have the need or right to see customers' data. Data access needs to be logged and tracked for compliance.
5. *Monitor your cloud environment for security threats*: Information security practices focus on defending systems against known threats. The attack surface in a public cloud is fairly large especially with multi-tenanted systems. Active monitoring is required to see who is accessing which data. To enforce partitioning, the strategy is to use temporal rotating keys [15], which are hard to compromise.
6. *Perform routine penetration tests*: Even while hardware and software in a public cloud are deemed secure, the combination of these may have a vulnerability due to the security assumptions made by one part about the other. This can only be uncovered by conducting end-to-end penetration test [16]. Performing these tests on a regular basis helps to identify gaps well before a breach occurs.

7. *Train employees on cloud security practices*: Often the biggest security threat in a public cloud is an enterprise and its employees. This is due to an inadvertent misuse of cloud environment, such as mishandling of security keys, or leaving critical data encrypted or including a malware in the software picked up from the public domain, etc. Thus, one should take time to train employees who will be using the public cloud system or working on an internal cloud that is open to the public.

6.7 Role of ML for Cybersecurity

Machine learning (ML) offers interesting opportunities to discover and plug-in vulnerabilities. Note that bad actors can also use ML to identity vulnerabilities and attack. ML has been very successful with complex tasks such as image recognition and natural language processing. Now it is being applied to improve cybersecurity [17]. This problem is more complex as human attackers strive to compromise the security of a system. So, the nature of problem keeps changing dynamically. Let us start by examining some newly emerging applications that traditional security algorithms do not address satisfactorily.

Imagine a scenario where multiple entities wish to come together for a common purpose, but do not quite trust each other. An example is new drug research that needs hospitals to provide patient data, pharmaceuticals to provide their drug data, and medical researchers to explore new treatment protocols. Neither party may want to give away its crown jewels but all are interested to know which drug protocols are effective in fighting and preventing diseases. For this type of situations, multi-party clouds offer a viable solution. There multiple users come together on a shared hardware to accomplish a common computational goal. It allows data of each party to be kept secured from other users.

As shown in Fig. 6.4, each user can start sending private data for computations, after being authenticated into the system. Proxy server hides the traceability of messages sent from each user toward the cloud server. This data is encrypted to provide protection and integrity against a man-in-the-middle attack. Analyzer in this model is the external party, which receives the statistical parameters of user data, decrypts it, and performs the analytics on it. There are obvious questions on the performance and efficiency of cloud environments to deploy such a model, while enforcing the security requirements of all user parties.

Privacy preserving algorithms for data mining tasks, which look for trends and hidden patterns, use techniques such as clustering, classification, or associate rule mining. Following are some of the strategies to improve security assurance in cloud operations:

1. *Anonymization*: This approach partitions a database table in two subgroups, containing Personally Identifiable Information (PII) and the rest. It removes the PII set of personal information, such as a person's name, date of birth, etc. These attributes are replaced by an assigned ID for the purpose of performing analytics.

Fig. 6.4 A secure multi-party-based cloud computing framework [18]

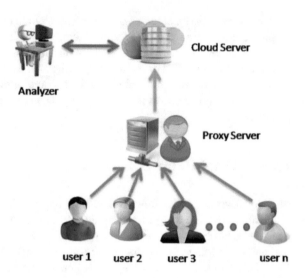

However, an attacker can use some prior or external information, such age, gender, and zip codes to decipher the identity of subjects. An example is de-anonymization of Netflix dataset based on the published movie ratings, and externally available datasets to identify a subscriber's records in a large dataset [19]. An earlier example of this is an interval classifier for database mining applications [20]. In this case, given a large population database that contains information about population instances. Given a sample, which is much smaller than the population, but representative of it, this classifier can retrieve all instances of the specified group from the larger population. Thus, anonymization is not always an effective technique.

2. *Secure Multi-Party Computation (SMPC)*: This strategy [18] considers all attributes of a dataset as private and uses cryptographic protocols for peer-to-peer communications. These techniques tend to be secure but also slow, so they do not scale well for large datasets.

3. *Randomization*: This approach seeks the underlying data while preserving the statistical properties of the overall dataset. Examples are additive data perturbation, random subspace projections, or simple mixing of time series from different sensors within acceptable windows or slices of measurement intervals. These techniques are fast and efficient, but do not provide security guarantees.

Now we shall examine how ML techniques are used to solve security problems. A common thread to all of these is to detect and learn from the past trends, as indicated below:

1. *Supervised learning*: It can be accomplished by using examples of executable files and labeling them as malware or harmless. Based on this training dataset, a model can learn and make decision about the new files. A disadvantage is the limits of labeled data, i.e., if a new data type differs from the training data, then it will not be classified correctly during the prediction phase.

2. *Ensemble learning*: By using different simple supervised models and combining them, a more sophisticated model can emerge. An example [21] is rating a movie. If a producer creates a new film and wants to get honest feedback on it, then may show it to more than one person. However, showing it 5 or even 50 people may not be sufficient if they all come from the same background. Thus, getting a feedback from a diverse set of audience is more likely to predict the final outcome of a movie's public release.

3. *Unsupervised learning*: If there is no labeled data, then model can learn itself by clustering. It can do so by detecting patterns or anomalies in the datasets. Currently, this works less precisely than the supervised approaches. Clustering is an example of unsupervised learning, e.g., when a professor wants to determine where to draw the cut-off lines for different grades in a class. She may look for the grouping of scores with gaps between them to differentiate A vs. B grade students. A key distinction between supervised and unsupervised learning is the access to prior labels or lack thereof.

4. *Semi-supervised learning (SSL)*: It combines benefits of both supervised and unsupervised approaches, when only some labeled data is available. Labeling audio files is a resource-intensive task [22]. Using a few labeled speech samples, and a large number of unlabeled datasets, SSL offers an excellent technique to generate accurate analysis of large speech samples.

5. *Reinforcement learning*: This is a trial and error process, where the model reacts to the outcome of its own decisions. Using reinforcement learning [23], Google was able to reduce energy consumption in its datacenters.

6. *Active learning*: This is similar to the reinforcement learning, except when the environment is also changing. An example is to look for terrorist threat by combing through social media posts.

6.8 Summary

Cloud computing exacerbates computer security issues arising from multi-tenancy and open access due to multiple users sharing resources. However, an even bigger challenge to information security has been created with the implementation of cloud computing. This chapter gave a brief general description of information security issues and solutions. Some information security challenges that are specific to cloud computing have been described. Security solutions must offer a tradeoff between the extent of security and the level of performance cost. A proposition of this chapter is that security solutions applied to cloud computing must span multiple levels and across functions.

6.9   Points to Ponder

1. Why is it important to separate out private data for encryption?
2. How does a public cloud make security issues more complicated than an enterprise or private cloud?
3. What is the need to encrypt data during transmission from a user site to the public cloud data center?
4. Are there any runtime concerns for data and programs in a public cloud?
5. Traditional AI has two distinct phases for training and inference. How does this need to change for security applications?

References

1. Sehgal, N. K., Bhatt Pramod Chandra, P., & Acken, J. M. (2019). *Cloud computing with security*. New York: Springer. https://rd.springer.com/book/10.1007/978-3-030-24612-9.
2. https://www.csoonline.com/article/3444488/equifax-data-breach-faq-what-happened-who-was-affected-what-was-the-impact.html.
3. Sehgal, N., Xiong, Y., Mulia, W., Sohoni, S., Fritz, D., & Acken, J. (2011). A cross section of the issues and research activities related to both information security and cloud computing. *IETE Technical Review, 28*, 279. https://doi.org/10.4103/0256-4602.83549.
4. Enck, W., Butler, K., Richardson, T., McDaniel, P., & Smith, A. (2008). Defending against attacks on main memory persistence. In *Proceedings of the 2008 annual computer security applications conference* (pp. 65–74).
5. Cachin, C., Keidar, I., & Shraer, A. (2009). Trusting the cloud. *SIGACT News, 40*, 81–86.
6. Naor, M., & Rothblum, G. N. (2009). The complexity of online memory checking. *Journal of the ACM, 56*, 1–46.
7. Juels, A., & Kaliski, B. S., Jr. (2007). PORS: Proofs of retrievability for large files. In *Proceedings of the 14th ACM conference on computer and communications security, Alexandria, VA, USA* (pp. 584–597).
8. Christodorescu, M., Sailer, R., Schales, D. L., Sgandurra, D., & Zamboni, D. (2009). Cloud security is not (just) virtualization security: A short paper. In *Proceedings of the 2009 ACM workshop on Cloud computing security, Chicago, IL, USA* (pp. 97–102).
9. Moscibroda, T., & Mutlu, O. (2007). Memory performance attacks: Denial of memory Service in multi-core systems. In *Proceedings of 16th USENIX security symposium on USENIX security symposium, Boston, MA* (pp. 1–18).
10. Saripalli, P., & Walters, B. (2010). QUIRC: A quantitative impact and risk assessment framework for cloud security. In *2010 IEEE 3rd international conference on cloud computing (CLOUD)* (pp. 280–288).
11. Ristenpart, T., Tromer, E., Shacham, H., & Savage, S. (2009). Hey, you, get off of my cloud: Exploring information leakage in third-party compute clouds. In *Proceedings of the 16th ACM conference on computer and communications security, Chicago, IL, USA* (pp. 199–212).
12. Osvik, D., Shamir, A., & Tromer, E. Cache attacks and countermeasures: The case of AES. In D. Pointcheval (Ed.), *Topics in cryptology—CT-RSA 2006* (Vol. 3860, 2006, pp. 1–20). Berlin/Heidelberg: Springer.
13. https://solutionsreview.com/Cloud-platforms/7-Cloud-security-best-practices-to-keep-your-Cloud-environment-secure/.
14. https://www.cio.com/article/3273108/understanding-the-benefits-of-a-multi-Cloud-strategy.html.

15. Everspaugh, A., Paterson, K., Ristenpart, T., & Scott, S. (2017). Key rotation for authenticated encryption. https://eprint.iacr.org/2017/527.pdf.
16. https://en.wikipedia.org/wiki/Penetration_test.
17. https://towardsdatascience.com/machine-learning-for-cybersecurity-101-7822b802790b.
18. Pussewalage, H. S. G., Ranaweera, P. S., Oleshchuk, V. A., & Balapuwaduge, I. A. M. (2013). Secure multi-party based Cloud Computing framework for statistical data analysis of encrypted data. *ICIN 2016, at Paris.* http://dl.ifip.org/db/conf/icin/icin2016/1570221695.pdf.
19. https://www.cs.cornell.edu/~shmat/shmat_oak08netflix.pdf.
20. Agrawal, R., Ghosh, S., Imielinski, T., Iyer, B., & Swami, A. (1992, August). An interval classifier for database mining applications. In Proceedings of the VLDB conference (pp. 560–573). http://citeseerx.ist.psu.edu/viewdoc/download?doi=10.1.1.50.98&rep=rep1&type=pdf.
21. https://www.analyticsvidhya.com/blog/2018/06/comprehensive-guide-for-ensemble-models/.
22. https://medium.com/@jrodthoughts/understanding-semi-supervised-learning-a6437c070c87.
23. https://www.forbes.com/sites/bernardmarr/2018/09/28/artificial-intelligence-what-is-reinforcement-learning-a-simple-explanation-practical-examples/#2561c00139ce.

Chapter 7
Examples of Analytics in the Cloud

7.1 Background

A vast amount of information is available on the Internet. Also, it is growing at an overwhelming rate. Individual users and businesses are struggling to store and use this information in meaningful ways. Solutions are emerging in the way of a new class of algorithms, such as MapReduce. We shall study it in this chapter. Let us first consider the statistics from 2019 about what people do on the Internet in a typical minute [1]. This is depicted in Fig. 7.1.

Google is able to process 3.8 million search queries in a minute. Online users are sending 188 million emails every 60 s, undoubtedly many of these from online systems ending up in spam folders. In the previous year of 2018, only 67 voice-activated devices were being shipped per minute, which has jumped to 180 smart speakers in 2019. Meanwhile, Netflix has tripled the number of people watching its content, and Instagram has double the number of users scrolling through feeds on their devices.

As staggering as these statistics seem, the background story is even more impressive. For every 100 new smartphones sold, a new hardware server is required in the datacenter to support the new users and applications. All these activities are generating mountains of new data every second that business need to process for better decision-making.

7.2 Analytics Services in the Cloud

Analytics Services in the Cloud refers to the process of using Cloud Computing resources to process large amounts of data [2]. It uses a range of analytical tools and techniques to help businesses extract useful information from massive datasets. The results are presented in a report or via a Web browser. Due to the elastic nature of

© The Author(s), under exclusive license to Springer Nature Switzerland AG 2021
P. Gupta, N. K. Sehgal, *Introduction to Machine Learning in the Cloud with Python*,
https://doi.org/10.1007/978-3-030-71270-9_7

Fig. 7.1 Activities taking place on Internet in a typical minute [1]

Cloud Computing, and variable nature of resources required for AI/ML tasks, it is economically beneficial to use cloud instead of on-premise installed servers.

Below are some examples of analytics products currently offered by the big three public Cloud Service Providers (CSPs) in the United States:

1. *Amazon Web Services (AWS)*:

    (a) *Athena*: It is an interactive query service to analyze data in Amazon S3 (storage repositories) using standard SQL. Athena is serverless, so there is no infrastructure to manage, and user only pays for the queries execution time [3]. Athena works directly with the data stored in AWS's Simple Storage System (S3), using a distributed SQL engine Presto to run the queries. It

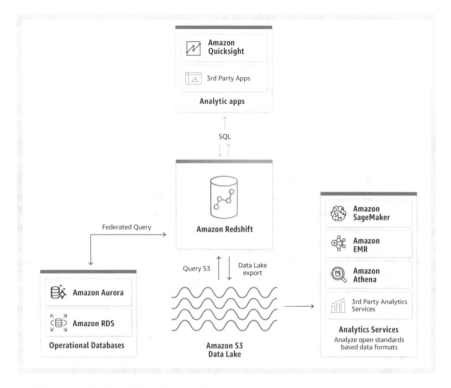

Fig. 7.2   Amazon's Cloud Stack for analytics on large datasets [7]

works with a variety of standard formats such as CSV, JSON, ORC, Avro, and Parquet. It can also handle complex analysis, including large joins, window functions, and arrays. Athena uses an approach known as schema-on-read, which overlays a schema on the data at the time of a query's execution. It uses Apache Hive to create, drop, and alter tables and partitions [4].

(b) *Elastic Map Reduce (EMR)*: It is a cloud-native big data platform for processing vast amounts of data at scale [5]. The underlying engines use opensource tools such as Hadoop, Spark, and Hive [4, 6].

(c) *Redshift*: It is a fully managed, petabyte-scale data warehouse to run complex queries on collections of structured data [7]. The underlying technology uses Massive Parallel Processing (MPP) data warehouse to handle large-scale datasets and database migrations, as shown in Fig. 7.2. The figure illustrates the ability for third-party applications to access operational data stored in relational databases. Redshift enables applications to query data, and write it back to the data lake in open formats. This offers an option to store highly structured, frequently accessed data in a Redshift data warehouse, while also keeping exabytes (10^{18} bytes or 1 billion gigabytes) of semi-structured or unstructured data in S3. The lowest layer of the architectural stack shows how exporting data from Redshift to data lake enables analyzing it further with AWS services such as Athena, EMR, or SageMaker. Use-cases include business intelligence (BI) tools for operational and descriptive analytics on real-time business events, such as generating quarterly results.

2. *Google Cloud Platforms (GCP)*: Google offers a suite of cloud computing services [8] that use the same infrastructure as Google uses internally for its end-user products, such as search and YouTube. In addition to providing computing services on demand, Google offers data storage, analytics, and machine learning through GCP.

   (a) *BigQuery*: It is a RESTful web service that enables interactive analysis of massive datasets. If offers a fully managed, low-cost analytics data warehouse, which can be used with Software as a Service (SaaS) applications [9].
   (b) *Dataproc*: It provides Spark and Hadoop services [4], to process big datasets using the open tools in the Apache big data ecosystem.
   (c) *Composer*: It is a fully managed workflow orchestration service to author, schedule, and monitor pipelines that span across clouds and on-premise datacenters.
   (d) *Datalab*: It offers an interactive notebook (based on Jupyter) to explore, collaborate, analyze, and visualize data using Python.
   (e) *Studio*: It takes data into dashboards and reports that can be read, shared, and customized across users.

3. *Microsoft Azure*: It is an open, flexible, and enterprise grade cloud computing platform [10]. Microsoft Azure offers cloud computing services through its data centers for building, testing, deploying, and managing third-party applications. These are hosted in virtual machines running in Linux or Windows operating environments.

   (a) *HDInsight*: It supports the last open source projects from the Apache Hadoop and Spark ecosystems [1]. In addition, HDInsight provides data protection with monitoring, virtual networks, encryption, Active Directory authentication, authorization, and role-based access control. Multiple languages and tools such as Python, Jupyter Notebook, and Visual Studio are supported.
   (b) *Data Lake Analytics*: It provides distributed analytics service for using and managing big data. This is done through U-SQL, an extensible language that allows code to be parallelized at scale for query and analytics.
   (c) *Machine Learning Studio*: As the name suggest, it enables users to build, deploy, and manage predictive analytics solutions.

An example of Azure data flow is shown in Fig. 7.3. It enables Cloud customers to run massively parallel jobs. Azure data factory converts over petabytes of raw data coming in from multiple sources to be stored in Azure Data Lake Store. Then it can be queried using SQL data warehouse, to generate actionable business insights [11]. Following steps are followed:

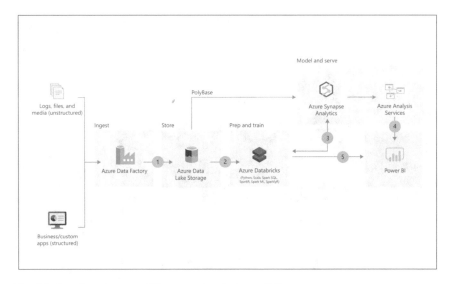

Fig. 7.3 Data flows between different services in Azure [11]

1. Raw data is ingested from multiple sources and stored in Azure Data Lake Storage.
2. Using Python, Scala, and Spark SQL to cleanup and prepare this data for AI.
3. Then an AI model is trained and built using Azure Analysis services.
4. Lastly, Azure's power BI tools are used to extract business metrics from this data.
5. Any model tuning and corrections are done comparing predictions with the expected values.

7.3   Introduction to MapReduce

MapReduce is a programming model for processing and generating big datasets using distributed, parallel algorithms on a cluster of servers [11]. As its name indicates, a MapReduce program has two steps: a map procedure to perform filtering and sorting, followed by a reduce method to perform a summary operation. An example of map is to sort the students by their last names into a queue, one queue for each name. Example of a reduce is to count the number in each queue, yielding last name frequencies. Another example, as shown in Fig. 7.4, illustrates a program to compute the frequency of different alphabets in the given strings. In the split phase, each word is mapped to a different processor, which in the map phase does the parsing and computation of each alphabet in a given string. Then in the combine phase, each alphabet's occurrence is aggregated. These individual counts are finally reduced to yield the desired results.

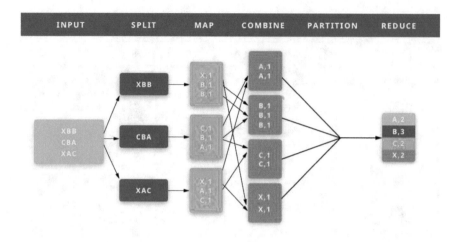

Fig. 7.4 An example of MapReduce Framework to count the alphabet frequency [6]

An advantage of running these tasks in parallel is to gain performance, as well as provide redundancy and fault tolerance. The latter is obtained by creating multiple copies of the same data on different disks, then fetching it in parallel, comparing the results to decide the correct value. Thus, MapReduce is a good solution for achieving distributed parallelism among commodity machines, However, a single-threaded implementation of MapReduce is usually not faster than the traditional computation, but a multi-threaded implementation on a multi-processing hardware is faster. Google uses it for Wordcount, Adwords, Pagerank, and Indexing data. Facebook uses it for various operations including analytics and demographics classification of its large user base.

7.4 Introduction to Hadoop

A popular open-source implementation with support for distributed shuffles can be found in Apache Hadoop [4]. It is a collection of software utilities that use a network of many computers to solve problems involving massive amounts of computation and data. The core of Apache Hadoop consists of a storage part, known as Hadoop Distributed File System (HDFS), and a processing part that is the MapReduce programming model, as shown in Fig. 7.5.

Hadoop splits given files into large blocks and distributes them across nodes in a cluster of servers. It then transfers packaged code into nodes to process the partitioned data in parallel. This approach benefits from the data locality, where server nodes operate on the mapped data that they have direct access to. This enables a large dataset to be processed faster and more efficiently than it could in a

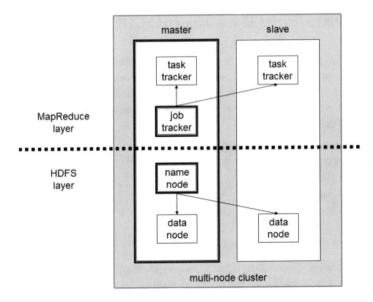

Fig. 7.5   A multi-node Hadoop Cluster [4]

conventional computing environment. Base of Apache Hadoop framework is composed of the following modules:

1. *Hadoop Common*: It contains libraries and utilities needed by other Hadoop modules.
2. *HDFS*: It is a distributed file system that stores data on regular machines, yielding a very high aggregate throughput across the cluster.
3. *Yarn*: It is a platform responsible for managing computing resources in clusters and using them for scheduling end users' applications.
4. *MapReduce*: It is an open-source implementation of the MapReduce programming mode [21] for large-scale data processing.

In a larger cluster, HDFS nodes are managed through a dedicated NameNode server to host the file system index, and a secondary NameNode that can generate snapshots of the NameNode's memory structures, thereby preventing file system corruption and loss of data.

7.5   Examples of Cloud-Based ML

In this section, we will look at some use cases and examples of deploying AI and ML solutions in the Cloud.

7.5.1 Cloud Security Monitoring Using AWS

Security in a public cloud is a hotly debated topic. One of the concerns is unauthorized access leading to loss of data. A proposed solution [12] uses AWS CloudTrail and CloudWatch logs, which are stored and mined for detecting suspicious activities. Goal of such security monitoring is to mitigate one or more of the following risks:

- Weak identity or user credentials
- Insecure APIs
- Account hijacking
- Malicious insiders
- Advances Persistent Threats (APTs)
- Data loss
- Abuse or nefarious use of cloud services

AWS already provides Identity and Access Management (IAM) services. It has options available to configure for different level of access permissions. Following the principle of least privilege, it is recommended to give least amounts of permissions to manage AWS resources required to perform the intended job function. ML is ideal for controlling various AWS credentials, since it can learn from the previous events, establish a normal pattern and identify anomalies. The proposed method [12] uses Supervised Learning technique with a linear regression to predict risk scores for AWS Cloud infrastructure events. An experimental setup is shown in Fig. 7.6. Splunk is used to ingest AWS Cloud trail and CloudWatch logs to implement security monitoring. The steps involved are as follows:

1. Collect all of the AWS log data to Splunk
2. Visualize and combine data with filters
3. Apply ML models to build baselines
4. Develop risk scores using ANN, instead of manual rules/thresholds
5. Choose methods for estimating model performance
6. Evaluate results, tune the parameters, and deploy the model
7. Identify any suspicious access attempts in AWS infrastructure

The types of events analyzed [12] were as follows, and the system is able to detect a suspicious activity and assign a risk scores as shown in Fig. 7.6.

- aws_cloudtrail_notable_network_events
- aws_cloudtrail_iam_change
- aws_cloudtrail_errors
- aws_cloudtrail_change
- aws_cloudtrail_delete_events
- aws_cloudwatch_sns_events
- aws_cloudtrail_auth
- aws_cloudtrail_iam_events
- aws_cloudtrail_ec2_events

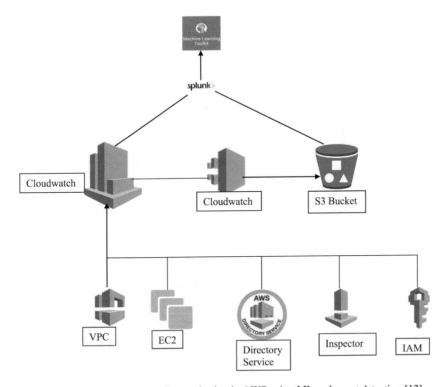

Fig. 7.6 Proposed setup for security monitoring in AWS using ML and event detection [12]

7.5.2   *Greener Energy Future with ML in GCP*

To reduce dependence on fossil fuels, natural renewable resources such as solar or wind power is used to generate electricity. AES operates large farms with hundreds of wind turbines on single tract of land. Inspect their structural integrity and any cracks in the blades can take several weeks of dangerous climbing. Instead of manual inspections, Measure [13] uses drones for collecting visual data within in a few hours. These data and images are then processed in Google's GCP cloud to analyze for defects. This enables humans to be involved in repair or replacement only as needed.

As shown in Fig. 7.7, drones from Measure collect about 300 images per turbine. These are processed and analyzed as follows:

1. Measure provides raw data sets and annotated inspection images from past turbine inspections.
2. This is used to train Google's computer vision models.
3. Measure's human experts then validate and refine the models.
4. New data from field is provided to AI/ML to identify turbine defects.
5. Engineers then visit the defective turbines for corrective actions.

Fig. 7.7 Drone-based inspection of wind turbines [13]

With ML, images with and without defects are identified and sorted automatically. This vastly reduces the time between the data collection and inspection results delivery. ML algorithms also classify the severity of defects, so experts only need to examine a few images with specified defects and recommend the best course of action.

7.5.3 Monorail Monitoring in Azure

Microsoft has worked with Scomi Engineering Bhd, a monorail manufacturer based in Malaysia, to develop a proof of concept (PoC) [14] for monitoring and predicting future maintenance needs by using Azure Machine Learning. These monorails are

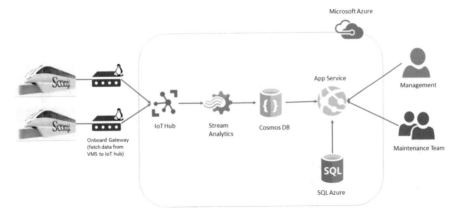

Fig. 7.8 An IoT setup for monitoring and maintaining monorails [14]

deployed in public transportation systems in several countries, including Malaysia, India, and Brazil. There is a need to collect telemetries from these trains into a centralized environment. Currently, the train status and alerts are sent manually to a central control center, enhancing the possibility of miscommunication and delays in rectifying issues. This may also result in downtime for the service and may even lead to an accident.

Goals set for this PoC were to:

1. Pull data from a train's Vehicle Management System (VMS).
2. Store it in the Cloud and process it for any alerts.
3. Present the results via a web interface.

Scomi decided to use Microsoft's Azure IoT Hub with a specialized gateway to push data from a train's controller to the cloud. Communication between an onboard Linux controller and IoT Hub is authenticated by using a device key registered in Azure Cloud. HTTPS protocol ensured security and frequency of data extraction was once every 5 s. In actual production environment, it can be set as often as every second. As shown in Fig. 7.8, data is processed through Azure Stream Analytics for near real-time analysis. It is then stored as schema-free in Azure Cosmos DB, using JavaScript Object Notation (JSON). Purpose of this repository is to perform offline machine learning activities. Access to Cosmos DB is also authenticated by using a key that is generated in Azure Cosmos DB settings. Web application access is secured by using ASP.NET identity, which provides claim-based access control. This is akin to security token service.

Device messages are sent every 5 s, and resulting data formatted in JSON appears as shown in Fig. 7.9.

For Stream Analytics, output is split into two. One is for storing the raw data in the Cosmos DB collection for future machine learning and analysis. The other is for filtering out alarm data and alerting the train operators. Former is also used for automating the task assignments sent to the maintenance department. They can use it for predictive maintenance of the monorail trains.

```
MsgSender = D2CMsgSender('HostName={iothubname}.azure-
devices.net;SharedAccessKeyName=iothubowner;SharedAccessKey={SharedKey}')
print(MsgSender.sendD2CMsg('RSV21',msgstring))
```

```
[
    {
        "trainid":"RSV 21",
        "timestamp":"2017-05-24 16:26:16",
        "id":44
    },
    {
        "trainid":"RSV 21",
        "timestamp":"2017-05-24 16:26:16",
        "id":1121
    },
    {
        "trainid":"RSV 21",
        "timestamp":"2017-05-24 16:26:16",
        "id":1121
    },
    {
        "trainid":"RSV 21",
        "timestamp":"2017-05-24 16:26:16",
        "id":1121
    },
    {
        "trainid":"RSV 21",
        "timestamp":"2017-05-24 16:26:16",
        "id":1121
    },
    {
```

Fig. 7.9 Train data packets pushed appear as JSON in Azure IoT Hub [14]

```
        "trainid":"RSV 21",
        "timestamp":"2017-05-24 16:26:16",
        "id":1121
    },
    {
        "trainid":"RSV 21",
        "timestamp":"2017-05-24 16:26:16",
        "id":1121
    },
    {
        "trainid":"RSV 21",
        "timestamp":"2017-05-24 16:26:16",
        "id":3156
    },
    {
        "trainid":"RSV 21",
        "timestamp":"2017-05-24 16:26:16",
        "id":3156
    },
    {
        "trainid":"RSV 21",
        "timestamp":"2017-05-24 16:26:16",
        "id":3156
    },
    {
        "trainid":"RSV 21",
        "timestamp":"2017-05-24 16:26:16",
        "id":3156
    }
]
```

Fig. 7.9 (continued)

7.5.4   Detecting Online Hate Speech Using NLP

Natural Language Processing (NLP) is one of the recent advances due to AI/ML techniques, mapped to cloud computing. This has been used to detect hateful words in online posts, which is an important problem to solve. There was a recent project, done as an undergraduate senior project by Hetesh Sehgal in Santa Clara University [15].

In today's world, more than half of Americans (53%) claim that they were subjected to hateful speech or online harassment in 2018 [16]. Additionally, studies show that an increase in the hate speech on social media, which then leads to more crimes against the minorities in the physical world. Social media entities, such as Twitter, Facebook, and YouTube, attempt to hide vulgar comments using AI and ML. However, their effectiveness is limited as documented by *USA Today*'s report [16]. Unfortunately, hate speech is still easy to find on mainstream social media sites, including Twitter. Hetesh used 160,000 Wikipedia comments for data training and classified them in a range of 0–1 based on one of the following labels:

- Toxic
- Severely Toxic
- Obscene
- Threat
- Insult
- Identity Hate

K-fold cross validation was done by using 80% of the data for training and remaining 20% for testing. Thus, his program was tested using 153,000 comments that need to be given a value for each of the previous labels. Implementation used Logistics Regression, Multinomial Naïve Bayes, and an Artificial Neural Network (ANN).

For Logistic Regression, Hetesh utilized Term Frequency Inverse Document Frequency (TFIDF), which enabled him to gauge the importance of a word relative to its corpus. The more frequent a given word is used, the lower is its weight. This is because if a word is used more frequently, then it is more likely to be a common "normal" word. Therefore, it should be considered less harmful in its given context, whereas a word that has a rare occurrence can be assumed to be used for a special instance and needs special attention. An example of weight is shown in Fig. 7.10.

For Multinomial Naive Bayes, Hetesh utilized a count vectorizer. A count vectorizer is a simple way to tokenize a collection of documents and build one's own vocabulary of known words. Dependent on the amount of times that a word occurs in a given corpus, its weight is adjusted. For the neural network, Hetesh utilized a TFIDF vector, and similarly removed all stop words from the text. He then fitted training data to this vector. The results of Logistics Regression validation, shown in Table 7.1, confirm that there is no over-fitting.

Fig. 7.10 Weight computations for logistics regression [12]

$$w_{i,j} = tf_{i,j} \times \log\left(\frac{N}{df_i}\right)$$

$tf_{i,j}$ = number of occurrences of i in j
df_i = number of documents containing i
N = total number of documents

Table 7.1 Logistic
Regression validation results

Labels	Training AUC	Cross-Valid AUC
Toxic	0.964	0.957
Severe_Toxic	0.986	0.986
Obscene	0.984	0.982
Threat	0.986	0.967
Insult	0.974	0.966
Identity Hate	0.977	0.969

```
Train on 127656 samples, validate on 15958 samples
Epoch 1/3
 - 238s - loss: 0.1013 - val_loss: 0.0556
Epoch 2/3
 - 235s - loss: 0.0424 - val_loss: 0.0549
Epoch 3/3
 - 236s - loss: 0.0293 - val_loss: 0.0602
```

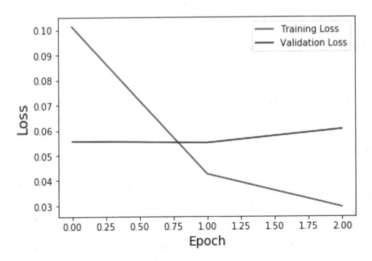

Fig. 7.11 Iterations to improve the neural network performance

Iterative training and validation resulted in improving the neural network parameters, as shown in Fig. 7.11.

The deployment of Hetesh's solution, as private score, and its comparison with
other public solutions using Kaggle website is shown in Fig. 7.12.

The above computations required a lot of computing power. This is economically
viable in public clouds. Furthermore, deployment of proposed solutions in online
media streams requires inference run-times to match the users' expectations of real-
time experience. A good solution should be able to detect and flag a writer upfront,
i.e., before the post is made available to readers, thus improving the overall online
experience.

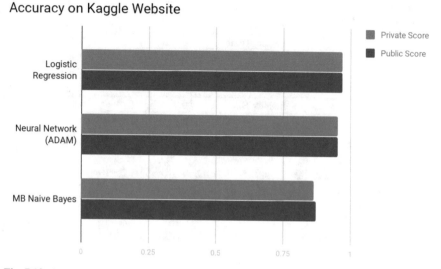

Fig. 7.12   Accuracy of AI/ML solutions for hate speech detection [17]

7.6   Future Possibilities

Since most real-world situations are dynamic, the corresponding dataset representations are also changing. Thus, training an AI/ML solution with a static, representative dataset will not sustain. This calls for reinforcement learning (RL), an area of machine learning concerned with taking actions in an environment to maximize cumulative rewards [18] with provisioning for dynamic adjustments. It differs from supervised learning, with a focus on finding a balance between the exploration of uncharted territory, and exploitation of current knowledge. Initial training is done as an offline process, but inference happens in the real-time most often in the cloud. It presents an opportunity to update the learning with feedback from local systems in the field back to the cloud-based AI/ML system. An example is AWS's SageMaker, where models can be trained without a large amount of data [19]. It is useful when the desired outcome is known, but the path to achieving it is unknown and requires many iterations to discover. Examples are healthcare treatments, optimizing manufacturing supply chains, and solving gaming challenges. Many of these are active research areas, e.g., cancer treatments for late diagnosed patients. Another example of RL is using Google Cloud to do parameter tuning [20]. It is helping to help advances fields such as self-driving cars, recommendation systems, and bidding. Training RL agents is expensive in terms of time and computing resources. It is not uncommon for an algorithm to take millions of training steps before the accumulated reward rises. Doing this in the cloud enables training for many models in parallel, evaluate different hyper parameters, quickly iterate, and converge on a solution. As Fig. 7.13 depicts, there is a feedback loop between the agent and the environment.

Fig. 7.13 A conceptual diagram of traditional reinforcement learning system

An agent may have two parts: one local and the other remote. The latter is typically located in the Cloud, directing an action in the field through the local agent on a device located in the environment. The sensor gives a feedback on the results to the remote agent, which can adjust its decision parameters to take corrective actions through the local agent as needed. It is important to have a local agent for immediate corrective actions. For example, braking in case of a self-driving car, which should not be relegated to the cloud, instead this decision should be taken by a local agent. However, the remote agent cloud can learn from many local agents as an offline process. This type of split decision-making is still an active research area. Therefore, we recommend a hybrid system with two agents, one local and the other remote, as shown in Fig. 7.14.

Both agents start with identical parameters with the initial training dataset. The training can happen in cloud as Step 1, with an action to set the weights of the Local Side Agent (LSA) in Step 2. However, as local side agent is exposed to the real-life environment with changing situations, some of these actions may result in non-optimal results in Steps 3 and 4, leading to an incremental learning curve. This updates the reward value and state of the training parameters, which are fed back to the Cloud Side Agent (CSA) in Steps 7 and 8, respectively. It may lead CSA to update its weights and convey it back to LSA in Step 9. This relationship may involve more than one instances of LSA, as in multiple self-driving car situations on different roads at the same time. Then it is up to CSA to integrate these inputs and issue a new set of instructions and actions in Step 9 to various LSAs.

Current state-of-the-art technology is not able to handle such situations, and definitely not in the real-time. If presented with an incrementally updated training dataset, most neural networks tend to recompute an entirely full new set of weights, instead of incrementally updating their existing weights. Thus, a small input change may result in disproportionately large change in the settings for AI/ML systems. A good research area is to enable incremental updates based on additional training data. Furthermore, it is desirable to split the problem into two parts, as depicted in Fig. 7.14 by using cloud-based reinforcement learning.

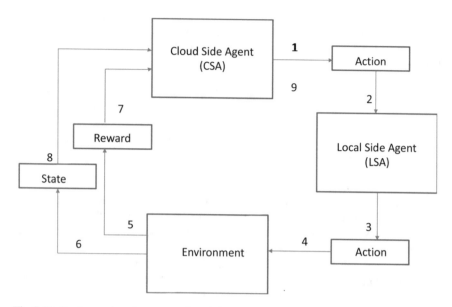

Fig. 7.14 Dual-agent iterative process for reinforced learning

7.7   Summary

Cloud with its vast reach and resources offers many tantalizing opportunities to improve AI-based learning systems. ML serves as a valuable tool for practitioners working on real-life problems using reinforcement learning. A key need is to develop the ability for enabling incremental training for neural network parameters, when the training dataset is constantly evolving. It is also an active research area with a promising future using cloud computing resources.

7.8   Points to Ponder

1. How has the spread of Internet enabled more data generation?
2. Why is it important to consider distributed decision-making in cloud?
3. Is data-based training a one-time activity or iterative?
4. Would Hadoop be still popular if it was not open-sourced?
5. How can the inference results of NLP feed back into the training phase?

References

1. https://www.visualcapitalist.com/what-happens-in-an-internet-minute-in-2019/.
2. https://en.wikipedia.org/wiki/Cloud_analytics.
3. https://aws.amazon.com/athena/.
4. https://en.wikipedia.org/wiki/Apache_Hadoop.
5. https://aws.amazon.com/emr/.
6. https://www.edupristine.com/blog/hadoop-mapreduce-framework.
7. https://aws.amazon.com/redshift/.
8. https://cloud.google.com/.
9. https://en.wikipedia.org/wiki/Google_Cloud_Platform.
10. https://azure.microsoft.com/en-us/.
11. https://blogs.msdn.microsoft.com/azurecat/2018/12/24/azure-data-architecture-guide-blog-8-data-warehousing/.
12. https://www.csiac.org/journal-article/cloud-security-monitoring-with-ai-ml-infused-technologies/.
13. https://www.measure.com/news-insights/measure-aes-and-google-automl-ai-wind-turbine-inspection.
14. https://microsoft.github.io/techcasestudies/iot/2017/07/25/ScomiEngineering.html.
15. http://hetesh.com/.
16. https://www.usatoday.com/story/news/2019/02/13/study-most-americans-have-been-targeted-hateful-speech-online/2846987002/.
17. https://www.kaggle.com/c/jigsaw-toxic-comment-classification-challenge/overview.
18. https://en.wikipedia.org/wiki/Reinforcement_learning.
19. https://aws.amazon.com/about-aws/whats-new/2018/11/amazon-sagemaker-announces-support-for-reinforcement-learning/.
20. https://cloud.google.com/blog/products/ai-machine-learning/deep-reinforcement-learning-on-gcp-using-hyperparameters-and-cloud-ml-engine-to-best-openai-gym-games.
21. https://en.wikipedia.org/wiki/MapReduce.

Chapter 8
Health Care in the Cloud: A Few Case Studies

8.1 Introduction

A key area of human development is health care. Within health care, an important consideration is to track basic indicators of human health. Thus far, there are four accepted medical signals that indicate whether a person is dead or alive, namely body temperature, heartbeat, oxygen saturation level, and blood pressure. However, someone in a coma or unconscious state, while hooked to ventilators in a hospital, will readily pass these four tests. There, an additional new signal is needed to test. Such a test ought to cover the quality of blood in a brain to better determine the well-being of a person. Brain is a muscle that weighs approximately 2% of the body weight, but consumes 20% of oxygen supply via blood flow in an active, healthy person. However, for someone who is brain dead, the blood supply will automatically reduce as the nature is an efficient engineer to conserve energy. Hence, by observing blood velocity and quality of flow in brain's main arteries, doctors can determine if a person has suffered stroke or hemorrhage, etc. Other tests to ascertain the same condition are CT scans or MRI. The former tends to expose a patient to excessive radiation, which if repeated can lead to tissue damage and even cancer. The latter is expensive and time consuming, leading to delayed treatment.

With this in mind, medical scientists have invented an ultrasound technique called Transcranial Doppler (TCD), which is deemed safer than X-rays, CT scans, or MRIs. TCD [1] is a noninvasive and painless ultrasound technique that uses sound waves to evaluate blood flow in and around the brain. No special contrast or radiation is involved in a TCD test. Physicians recommend this to determine if there is anything that is affecting blood flow in the brain. Hence, a new medical signal has been established to examine blood flow in the brain. It can add to the previously described four tests to ascertain well-being of a person. These observations led to formation of a new company called, NovaSignal, meaning a new signal. This company specializes in combining AI and ML, besides Robotics with the traditional TCD equipment to assist physicians treating brain diseases. According

P. Gupta, N. K. Sehgal, *Introduction to Machine Learning in the Cloud with Python*, https://doi.org/10.1007/978-3-030-71270-9_8

to NovaSignal's website [2], there are 800,000 strokes in US every year. Unfortunately, 90% of these stroke victims do not receive appropriate treatment within 6 h. Since time is critical for a successful treatment of stroke, a delay can lead to long-term health impairment for the patients. Furthermore, more than 50% of cardiac patients, who underwent aortic value surgery, also show imaging signs of strokes. This implies that a stroke-free patient may suffer from clots and brain damage, while recuperating in a hospital room after a successful heart surgery.

8.2 Existing TCD Solution

Since human brain is protected with a relatively thick bone—the skull, it is not possible for a low-energy ultrasound wave to penetrate it. However, mother nature has left few small openings where skull bone joins jaw and other bones in the body. These holes are used as windows to monitor brain, by positioning the ultrasound probes, as shown in Fig. 8.1. These signals are then reflected back by the brain.

The reflections vary in frequency following Doppler principle [3], similar to traffic police's use to measure speed of cars on a highway. If TCD Doppler waves encounter a nerve carrying cerebral (i.e., brain's) blood in a direction perpendicular to probe's surface, then the reflected waves will have a higher frequency or lower, depending on whether the blood flow is toward or away from the probe, respectively. Changes in frequency are proportional to velocity of the blood flow, as determined by Doppler's equations [3]. TCD signals travel through acoustic windows, which are the areas defined by the pathway of the ultrasound beam between the transducer and the acoustic reflectors, as shown in Fig. 8.2.

Fig. 8.1 Manual positioning of TCD probes [2]

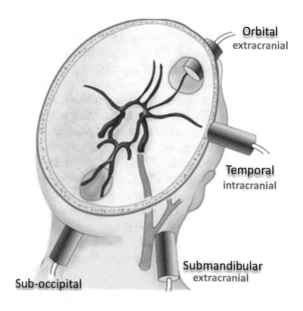

Orbital
extracranial

Temporal
intracranial

Submandibular
extracranial

Sub-occipital

Fig. 8.2 Ascertaining
blood flow in brain arteries
with TCD probes [2]

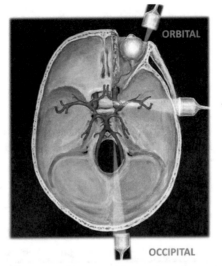

Fig. 8.3 Tracking brain's blood flow using NovaGuide's robotic probes

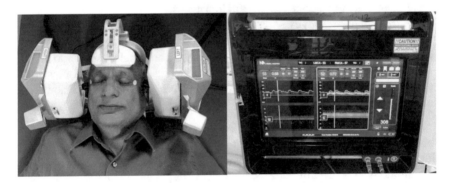

Without going into the medical details of probes' positioning and blood flow velocity measurements [4] through NovaSignal's equipment, we are interested in the data collection and how it can be shared with physicians in a location away from where TCD exam is conducted. Even before that, we must note that it is an art to position the probes correctly as the skull bone is broad and thick. Thus, an experienced sonographer is needed to conduct a TCD exam. This has been alleviated by NovaSignal's robots that can find optimal locations for probe automatic placements in a matter of few minutes or even sooner. It allows for unattended patient observations, e.g., someone unconscious or recovering in a hospital ICU, as depicted in Fig. 8.3.

Doctors have learnt while measuring rise and fall of blood in brain vessels, that it follows the pattern of heart beats. Both the height (indicating the minimum and maximum velocity of flow), and shape of the waveforms (slope indicates a potential stroke) are important. Detailed discussion of this is beyond the scope of this book, but can be found at [5–7].

8.3 Trail of Bubbles

Doctors at Mt. Sinai hospital are using NovaSignal's products to evaluate 13 major arteries supplying blood flow to the brain. Some examples of TCD studies [1] to identify specialized problems related to brain blood flow are as follows:

1. *Embolic detection monitoring*: This test detects any free-floating particles that may be in the bloodstream. These particles can be a significant source of stroke risk.
2. *Agitated saline bubble study*: This test is specific for identifying whether there is a passageway or hole between the right and left chambers through which blood enters the heart. This hole usually closes after birth. If it does not close, it can be a source of increased stroke risk in certain patients.
3. *CO_2 vasomotor reactivity study*: This noninvasive study looks at whether the small vessels that regulate blood flow to the brain are working properly. When a small amount of carbon dioxide (CO_2) and oxygen are inhaled, similar to holding one's breath, these small vessels should widen and increase blood flow to the brain. When a person hyperventilates, the vessels should shrink and slow blood flow to the brain.

A recent breakthrough came when Dr. Alex Reynolds, a Mt. Sinai Physician, checking comatose Covid-19 patients for signs of a stroke instead stumbled on a new clue about how the virus may harm lungs. This was due to air bubbles passing through the bloodstream of patients, who were not getting enough oxygen despite being on ventilators [8]. Since it is risky for health workers to be near the patients for long periods due to the nature of Covid spread, NovaSignal's robotic headset did automatic tracking once positioned on a patient, as was depicted in Fig. 8.3.

The result of Dr. Reynold's study using NovaGuide was detection of abnormally dilated lung capillaries, unrelated to a heart problem, which were letting the bubbles sneak through. At the end of a pilot study, 15 out of 18 tested patients had microbubbles detected in the brain. She showed that in some cases, ventilators were doing more harm than good. This study has opened new pathways to Covid patients' treatment around the world and shows the power of combining medical science with robotics.

8.4   Moving Data to the Cloud

NovaSignal's CEO, Diane Bryant, realized the need for remote patient monitoring and enabling Physicians to examine TCD datasets at a later time. She initiated a new team to develop cloud-based applications that can store and view TCD exams data away from the NovaGuide machines. The team used state-of-the-art cloud technologies, which ensure patient confidentially, data integrity, and high levels of application performance. An architectural diagram for deployment of NovaGuide View is shown in Fig. 8.4. It starts with NovaGuide's robotic TCD system in the yellow boxes at the top-middle. After patient was examined, data gets stored in a hospital's PACS (Picture Archiving and Communication System), from which the final report gets sent to EHR (Electronics Health Records). In parallel, the complete TCD exam data can be sent to a server inside the hospital, or to a server in the Public Cloud with necessary de-identification [9], as shown in the Fig. 8.4. This solution meets the existing HIPPA regulations [9].

8.5   A Reader in the Cloud

In order to support growing customers, there is a need to view TCD data away from a machine where patient was examined, a web-based app has been developed. It enables remote login from a PC, workstation, tablet, or even a phone by authorized users, e.g., Physicians Initial login screen is shown in Fig. 8.5.

Once login is successful, users have an option to search the current or past data by their type or date when conducted. Physicians can search the database in Cloud by the name or other HIPPA compliant particulars of a patient, as shown in Fig. 8.6 [9].

Fig. 8.4 NovaGuide View application in a Public Cloud

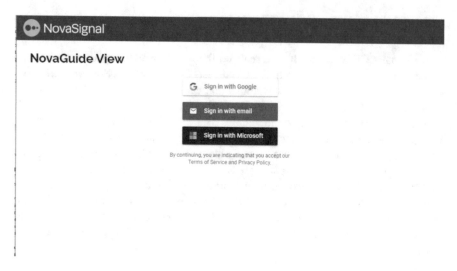

Fig. 8.5 Initial login screen for the Cloud App

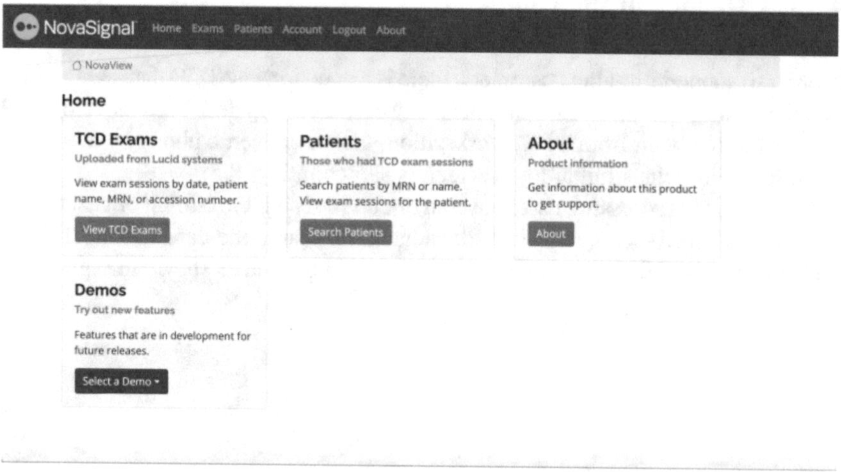

Fig. 8.6 Options to search by Exams or Patients [9]

Once the TCD data is secured in a private or public cloud, it can be accessed by physicians anytime, anywhere using a web-enabled application, as shown in Fig. 8.7. Doctors can also examine a patient's past examination data to assess change in the health conditions over time.

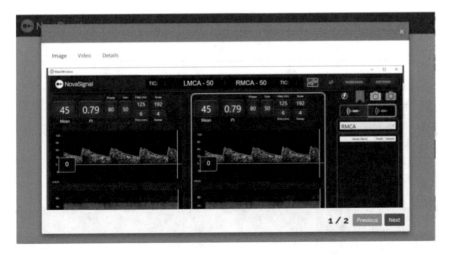

Fig. 8.7 Viewing a TCD exam on a remote screen [9]

8.6   Cloud-Based Collaborative Tools

Millions of people worldwide suffer from cancer. Once a diagnosis is done, painful regimen of treatment including radiation, chemotherapies, or surgeries begin. All of these tend to kill healthy cells along with the cancerous cells. If these patients can be treated as individuals based on their specific genome sequencing, and precision treatment can be drawn quickly, the course of disease can be significantly altered (Fig. 8.8).

With the above problem in view, Intel had launched a Collaborative Cancer Cloud (CCC) in 2015 that enabled institutions to securely share their patients' genomic, imaging, and clinical data for potentially lifesaving discoveries [10]. Such sharing allows large amounts of data from sites all around the world to be analyzed in a distributed system, while preserving the privacy and security of that patient The Collaborative Cancer Cloud is a precision medicine analytics platform that allows institutions to securely share patient genomic, imaging, and clinical data for potentially lifesaving discoveries. It will enable large amounts of data from sites all around the world to be analyzed in a distributed way, while preserving the privacy and security of each patient's data at each site.

Exploring solutions beyond cancer, a real problem lies in the fact that for a robust ML solution, we need to identify high-value groups of subjects for clinical trials, predict responses of patients to proposed treatments and pick relevant biomarkers, and extract new insights. Hence, a new technology has emerged to train ML models at scale across multiple medical institutions without moving the data between them. It is called federated learning (FL) [11], an instance of which is depicted in Fig. 8.9. FL enables data to stay local, and algorithms to travel across the participating institutions, for training a deep learning algorithm while preserving privacy and security of the patients' data.

COMBINING DATA:
COLLABORATIVE CANCER CLOUD

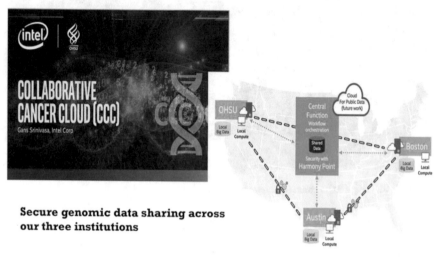

Fig. 8.8 Intel's Collaborative Cancer Cloud [10]

For example, Intel and University of Pennsylvania have announced a collaboration involving 29 international medical centers to train models to recognize brain tumors [12]. The goal of this is to provide diverse datasets for ML that no single institution can provide alone. This effort started recently and, if successful, will result in a new class of solutions that can identify brain tumors from a greatly expanded version of the International Brain Tumor Segmentation (BraTS) challenge dataset [13]. BraTS has been focusing on the evaluation of state-of-the-art methods for the segmentation of brain tumors in multimodal magnetic resonance imaging (MRI) scans. Adding FL-provided datasets to it will enhance the quality of results and may benefit patients around the world.

8.7   Multi-Cloud Solutions

Consider a company that develops medical research algorithms to aid doctors in making informed decisions. To leverage the cloud's storage and compute prowess, it needs to model and train algorithms based on hospital data. As hospitals adapt technology, they do so at various rates using various strategies to manage their own data. Let us consider three client hospitals, identified as Client1, Client2, and Client3. Client1 uses a local data storage, whereas Client2 uses Cloud A and Client3 uses Cloud B.

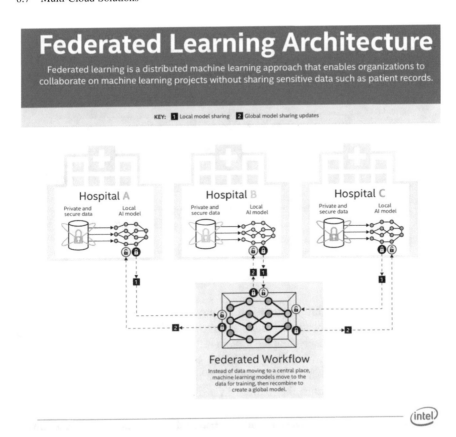

Fig. 8.9   A federated learning solution for ML and health care [12]

Suppose this company wants to run two types of algorithms: (1) standard computational software such as displaying measurement results in real time and (2) specialized algorithms that require further computations on the input datasets, which may not be yet approved for non-research usages. For standard practice computations, such as showing the results and output of a scan, required algorithms can be ported to cloud or local computer facility. However, for research algorithm deployments, the company may use a loosely coupled infrastructure to access data sources. This allows data to be pushed from a source, sent to a specific cloud where the research algorithms are hosted, and results may stay local or be pushed back to the cloud. This secure data transfer can be accomplished using a four-way handshake between APIs of two separate clouds or between a cloud and the local data warehouse. The diagram in Fig. 8.10 shows an example data that could be exchanged between Cloud1 (data source) and Cloud2 (application/computation cloud) for Client2 and Client3. It is important to note that these API layers can be configured at a broader level based on triggers for inputs and outputs. Specific source

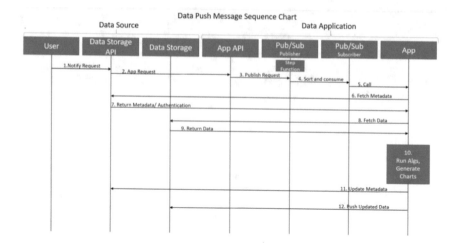

Fig. 8.10 A loose coupling between data source and applications [14]

and destination-related information can be passed at run time rather than having specific data values or locations hardcoded into a system.

An advantage of a loosely coupled system is that the data owner keeps full control and decide which algorithm they wish run on which dataset. Using message sequence chart [14], as shown in Fig. 8.10, a user on the left initiates services of an application or algorithm that resides in a different cloud on the right side. A request is sent to the application owner. Now it is up to the algorithm owner to honor the request. If it decides, then a counter request flows in the other direction as depicted by the right to left arrow in step 6. Note that dataset owner decides to give some minimal information via a metadata back to the algorithm owner, who then processes it and generates the required results. These are either sent back to the data owner or kept at the algorithmic source. The latter option is better if multiple data sources are needed for the computations, such as in the case of Collaborative Cancer Cloud [10].

8.8   UCSD Antibiogram: Using Unclassifiable Data

Computational medicine is a rapidly growing field. It attempts to apply machine learning techniques to massive medical datasets. It is especially popular with educational research organizations that have ready access to medical datasets. A particularly computationally intensive subdivision of this field is precision medicine, where scientists sequence a patient's genome and preempt future ailments. This may entail synthesizing personalized and even specifically crafted medications to treat a patient's condition. However, for the general case, attempting to build predictive models using medical data, especially in attempting to predict treatment efficacy on a given patient, yields low accuracy predictive models. This may present difficulties due to multitude of factors in the patient–treatment relationships. This is due to the way a single ailment can be caused or the high variability of the human immune response.

One particularly challenging issue within computational medicine is antibiotic selection. Antibiotics are life-saving drugs, but CDC (Center for Disease Control) estimates that one in three antibiotic prescriptions is unnecessary [15, 16]. Incorrectly prescribing antibiotics is one of the biggest factors contributing to the trend of microbial resistance, and antibiotic-resistant microbes are responsible for around 35,000 deaths in the United States annually [17]. The White House annually allocates a budget of over $1B toward combating antibiotic-resistant pathogens, and toward technology that promises to reduce their prevalence [18]. Clearly, design space has a huge financial and medical appeal.

Approaching this problem from a technical perspective, the first question is how much data is available. When a patient goes to the hospital with any infection, from strep throat to sepsis, samples of the bacteria are taken from the patient, and 5–20 antibiotics are tested on each sample to see which effectively kills the bacteria. Each of these tests are recorded as lab results including the concentration of antibiotic required to fully kill the bacteria, source of the infected sample, and tens of features regarding the test. This, in combination with the wealth of patient information recorded in the EHR (Electronics Health Records) containing demographics, ICD (International Classification of Diseases) codes, prescribed medication, and more, yields an extremely powerful and insightful dataset.

As stated, for every bacterial infection, samples are taken from the patient, and antibiotics are tested against those samples. Even after this exhaustive testing method, and subsequently prescribing the antibiotics that treated the bacteria effectively on a petri dish, an estimated one in three antibiotics are prescribed incorrectly [15, 16]. One reason for this complexity is associated with the variability of infectious diseases and human immune response.

Creating a viable solution for this medical problem is a large and competitive research space. One such research study is being conducted at UC San Diego. Researchers at UC San Diego obtained a dataset containing 5 years of antibiotic test information, related deidentified [18] demographic, social and hospital visit related data, and attempted to build an engine that would generate accurate antibiotic recommendations for a patient given various feature sets ranging from ~20 to over 100 features. To operationalize the dataset, researchers began with data scrubbing: from correcting spellings and recordings generated through the medical scription process. Next, the dataset was deduplicated according to digital antibiogram standards; extremely sick patients have longer hospital stays with many subsequent samples taken and antibiotic tests run inherently loading the dataset toward these special cases which need to be deduplicated to normalize the dataset. Finally, the dataset was operationalized in collaboration with medical professionals. This process involves creating medically tuned features from the available dataset by assessing medical truths from combinations of the infected sample site, ICD

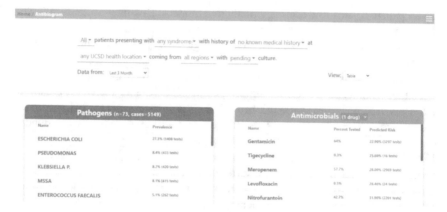

Fig. 8.11 GUI for UCSD's antibiogram application

codes, and various other less informative features, much like doctors do in their minds as they analyze patients.

After the dataset is ready for analysis, researchers attempt applying a multitude of learning techniques from classical (SVM, DT, PCA) to modern deep learning (NN, CNN), yet the results remained inconclusive, yielding accuracies too low for medical usage (85% accuracy minimum desired).

Sadly, this trend persists in many medical machine learning applications: high feature set, highly variable population and few direct causal relationships between lab results and presenting syndromes. Nonetheless, in UCSD study, the researchers were able to utilize power in their dataset by employing a Bayesian decision model. This network relies on the premise that although therapy efficacy cannot be universally classified on a highly varied patient population with high accuracy, there are many conditions that can heavily influence the outcome of a treatment. For example, patients with cystic fibrosis had especially low success with any antibiotic treatment due to an inherently weak immune system. By compounding the risk associated with each of these such factors, the dataset can be leveraged to demonstrate patient conditions that complicate antibiotic treatments (Fig. 8.11).

The resulting tool has been tested by UC San Diego clinicians. It allows them to enter specific details about a patient including their age, demographics, location of infection, previous infections, comorbidities, medical history, and many more critical factors. It displays an increasingly more accurately generated estimate of efficacy for each of the top recommended antibiotics and displays the remaining risk factors encountered by previous unsuccessful patients with similar medical histories.

This tool allows doctors to study treatment variability in specified patient cohorts and enables doctors to make highly informed decisions with more accurate predictions of patient outcomes. In spite of achieving low accuracy in a training/testing engine paradigm, allowing doctors to enhance their ability to study the variability of their patients and analyze key risk factors that they may not have been aware of because the dataset could be leveraged and yielded improvements in their medical process.

This project demonstrates a way to leverage data for integration into a process workflow: even with seemingly unclassifiable data, a dataset can create powerful and helpful solutions. This solution highlighted a need for codependence between the power of high-throughput analytical engines and the immense lexicon of medical knowledge available in medical professionals. Using this tool, UC San Diego doctors were able to critically analyze patient risk factors, how they influence the outcomes of various antibiotic treatments and improve the conditions of sick patients, all without the need for true classification.

There is always power in a dataset, it is up to the computational scientist to decide how it is best applied. In this case, when the medically required 85% accuracy was unable to be achieved, the dataset was still leveraged to improve the clinical process in a seamlessly integrated system that brings the dataset's power, enabling the medical professional to make the best possible decision.

8.9 Next Steps

Enormous other possibilities exist using cloud-based data repositories and ML tools. In the near future, using AI techniques described in this book, a prediction on the course of disease, along with a treatment recommendation, can be presented to doctors based on the training with labeled data from the past.

8.10 Summary

Health care may be the final frontier for humankind to win over new and existing diseases, which so far have not been conquered using last 101 years of technologies, as shown by the similarities between the 1918 and 2019 pandemics. This requires a perfect storm of medical data in the cloud with AI and ML solutions.

8.11 Points to Ponder

1. Discuss if we are at the cusp of a medical revolution with confluence of AI and cloud?
2. What are the advantages of TCD over MRI and CT scans for brain disease diagnosis?
3. Why some companies prefer a multi-cloud solution?

References

1. https://www.cedars-sinai.edu/Patients/Programs-and-Services/Imaging-Center/For-Patients/Exams-by-Procedure/Ultrasound/Transcranial-Doppler-TCD-ultrasound.aspx.
2. https://novasignal.com/.
3. https://en.wikipedia.org/wiki/Doppler_radar.
4. https://novasignal.com/wp-content/uploads/2020/07/DC01530-Rev-A-Basic-Technique.pdf.
5. https://www.frontiersin.org/articles/10.3389/fneur.2018.00847/full.
6. https://www.hcplive.com/view/this-advanced-ultrasound-headset-can-recognize-concussions-in-athletes.
7. https://www.frontiersin.org/articles/10.3389/fneur.2018.00200/full.
8. https://www.modernhealthcare.com/safety-quality/trail-bubbles-leads-scientists-new-coronavirus-clue.
9. https://www.hhs.gov/hipaa/index.html.
10. https://itpeernetwork.intel.com/intel-ohsu-announce-collaborative-cancer-Cloud-at-intel-developers-forum/.
11. https://owkin.com/federated-learning/.
12. https://www.hpcwire.com/2020/05/11/intel-upenn-launch-massive-multi-center-ai-effort-to-refine-brain-cancer-models/.
13. http://braintumorsegmentation.org/.
14. https://en.wikipedia.org/wiki/Message_sequence_chart.
15. CDC: 1 in 3 antibiotic prescriptions unnecessary. Centers for Disease Control and Prevention, Centers for Disease Control and Prevention, 1 Jan 2016. www.cdc.gov/media/releases/2016/p0503-unnecessary-prescriptions.html.
16. More people in the United States dying from antibiotic-resistant infections than previously estimated. Centers for Disease Control and Prevention, Centers for Disease Control and Prevention, 13 Nov 2019. www.cdc.gov/media/releases/2019/p1113-antibiotic-resistant.html.
17. Servick, K. White House plans big 2016 budget ask to fight antibiotic resistance. *Science*, (2017). www.sciencemag.org/news/2015/01/white-house-plans-big-2016-budget-ask-fight-antibiotic-resistance.
18. https://www.hopkinsmedicine.org/institutional_review_board/hipaa_research/de_identified_data.html.

Chapter 9
Trends in Hardware-Based AL and ML

9.1 Revisiting the History of AI

No surprises here that AI was in the top third searched terms on an IEEE site [1], just ahead of machine learning, with more readers interested in AI than Blockchain, Cloud computing, Internet of Things and 5G, combined. Top dozen most searched terms at the start of a new decade are shown in Fig. 9.1. This followed decades long AI winter cycles [2] starting in 1970s.

An AI winter refers to a period of reduced interest and funding cuts in the artificial intelligence research. It was coined by an analogy to the idea of a nuclear winter. However, AI winter was caused by a hype cycle, followed by disappointment and criticism. In a public debate at the annual meeting of American Association of Artificial Intelligence (AAAI), researchers Roger Schank and Marvin Minsky warned the business community that enthusiasm for AI had spiraled out of control in 1980s, and disappointment would certainly follow [1]. Within a few years, the then billion-dollar AI industry began to collapse. Recent years have witnessed a resurgence in the fields of AI and ML, led by a dramatic increase in funding and research. Most notable was DARPA's Grand Challenge Program, for fully automated road vehicles to navigate real-world terrains. As a result, many new applications and commercial projects using AI techniques are available in the market today.

A small sample is shown in the Fig. 9.2, as the underlying software platforms are becoming mainstream in the industry [3]. AI frameworks are evolving to ingest structured and unstructured data, for developing new predictive capabilities across all industries. However, most of these software capabilities are running on existing general-purpose hardware platforms, with traditional computational and memory access limitations. These companies, in order to distinguish themselves from their competition, have been seeking new custom hardware solutions for AI and ML.

© The Author(s), under exclusive license to Springer Nature Switzerland AG 2021
P. Gupta, N. K. Sehgal, *Introduction to Machine Learning in the Cloud with Python*,
https://doi.org/10.1007/978-3-030-71270-9_9

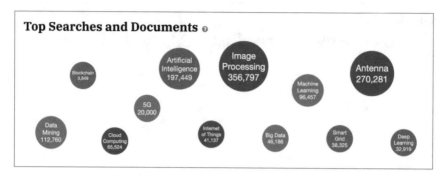

Fig. 9.1 Top search terms on IEEE digital library, as of January 2020 [1]

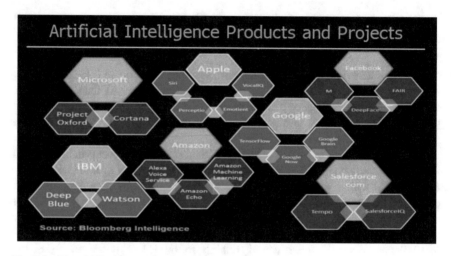

Fig. 9.2 AI- and ML-driven research by several market leaders [3]

9.2 Current Limitations of AI and ML

Although in theory, AI has an unlimited potential, but in practice it has several limitations due to the following factors:

1. *Lack of data*: If there is no sufficient training data, then neural network in an AI system cannot be trained properly.
2. *High model complexity*: To represent all the relationships in a given dataset, multiple dimensions may need to be modeled. This can exponentially increase the possible states in a model, and computations required to train the model.
3. *Available compute resources*: In order to train a model, and subsequently bring down the prediction errors of a previously trained model, there may not be sufficient CPU cycles or memory available.

Successful AI training and inference are dependent on the required computational availability, which in turn relies on energy. This requires electrical power to compute as well as air conditioning required to keep thermals under check. Besides power, field deployment of AI also depends on available form factors. For example, it is infeasible to pack a supercomputer in an autonomous car, even if the algorithms to drive are known. One way to partially solve #3, and help with #2, factors above is by eliminating several software layers of operating system, a virtual machine monitor, drivers, etc. This is possible by building special purpose hardware that can execute AI and ML algorithms. The speedup thus achieved will increase the available compute resources and also enable higher complexity models. Risk of not exploring such new avenues may lead to the possibility of another AI winter.

9.3   Emergence of AI Hardware Accelerators

An AI accelerator refers to a class of specialized hardware, designed to accelerate artificial intelligence applications. This includes neural networks, machine vision, and machine learning [4]. Such hardware may consist of multi-core designs, generally with a focus on low-precision arithmetic, novel dataflow architectures or in-memory computing capabilities [4–7]. AI accelerators can be found in many devices such as smartphones, tablets, and cloud servers. Their applications with limited success include natural language processing (NLP), image recognition and recommendation engines to assist in decision-making. These usage models gave rise to a new term, heterogeneous computing. It refers to incorporating a number of specialized processors in a single system, or even on a single chip. Each processor is optimized for a specific type of task, e.g., digital signal processing (DSP) for voice recognition or video gaming workloads.

9.3.1   Use of GPUs

The term GPU was coined by Sony in reference to PlayStation console's Toshiba designed graphics processing unit in 1994 [8]. Graphics processing units (GPUs) are specialized hardware for processing images [9], as depicted in Fig. 9.3.

Below is a brief description of units in the generic GPU shown in Fig. 9.3:

1. BIF is the bus interface functional unit, which connects the GPU to other components in a computer system. Its main role is to support bi-directional communications.
2. PMU is the power management unit. Its main role is to ensure correct supply voltages to various parts of the GPU. In addition, it can also be used as a throttling mechanism to control the frequency of the chip's clock.

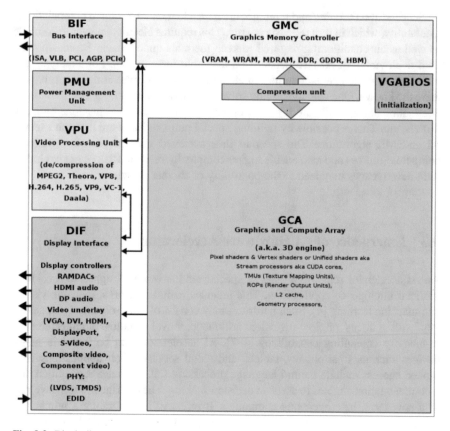

Fig. 9.3 Block diagram of a generic GPU [9]

3. VPU is the video processing unit. Its function is to do compression or de-compression of the bit-stream in different formats such as MPEG2.

4. DIF is the display interface unit. It has audio and video controllers for external devices that connect to the GPU, such as a screen monitor, or HDMI ports for a TV or a video projector.

5. GCA is the graphics and compute array. This is where a majority of matrix-based computations take place. It supports functions such as multiplication, rotation, translation done via geometric processors. This helps with computations for object rotation, translation, transposition, etc. There may be Level 1 and Level 2 caches located in this unit to optimize performance.

6. GMC is the graphics memory controller. It is needed as the entire dataset may not fit in the GCA registers, so read and write transactions happen via an internal bus between the compute engine and the memory storage units.

7. BIOS refers to basic input-output system and contains low-level software code to access the above hardware units. It is stored in the GPU to make it self-sustaining and accessible by other sub-systems in a computer server.

Such an architecture exploits data-level parallelism, by performing the same operation on multiple sets of data in a single step. GPUs exploit local image properties and often use single instruction multiple data (SIMD) type of parallel computing [10]. Images are digitally represented and stored in a system's memory as frame buffers for a display device. A GPU is able to rapidly operate on a system's memory in parallel steps. Their highly parallel structure makes them efficient for image processing.

With the emergence of new AI applications, relevance of GPUs has also increased. In recent research, it was found that computational genomics efficiency can be improved >200-fold in runtime, with five- to tenfold reductions in cost relative to CPUs [11]. This research uses open-source libraries such as Tensorflow and PyTorch, which were developed for ML applications using general purpose mathematical primitives and methods, e.g., matrix multiplication. A scale-up in the computations enables AI solutions involving complex models, larger datasets, resulting in more accurate predictions [12].

9.3.2 Use of FPGAs

Since deep learning frameworks are still evolving, many commercial applications do not warrant investment and design in a custom hardware. Field-programmable gate arrays (FPGAs) offer reconfigurable devices, which can run AI and ML algorithms at a faster speed than the general-purpose CPUs, or even GPUs in some cases. As an example, Microsoft reported use of FPGAs [13] to accelerate inference based on deep convolutional neural networks (CNNs). Its design showed performance throughput/watt significantly higher than for a GPU.

Amazon's AWS EC2 also provides a FPGA development kit to support high-performance compute instances [14]. Amazon F1 instances support Custom Logic (CL) designs, to create an Amazon FPGA image (AFI), as shown in Fig. 9.4.

The process to create an AWS F1 instance is as follows:

1. AWS provides a hardware development kit (HDK) along with a FPGA Amazon machine image (AMI). It contains a pre-built development environment, which includes scripts and tools for simulation of a design and compilation of its code.

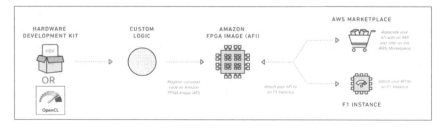

Fig. 9.4 Creation process for Amazon FPGA instances in cloud [14]

2. A developer can select this AMI and deploy it on an EC2 instance to provision the compute resources.
3. In addition, there is a GitHub repository available with AWS FPGA code samples.
4. AWS also provides cloud-based debug tools, such as a Virtual JTAG, Virtual LEDs, and DIP switches to emulate a physical development board.
5. The process starts by creating custom logic (CL) code. User writes the intended functionality in RTL (register transfer language).
6. Then, the CL is compiled using HDK scripts, which leverage Xilinx's Vivado tools to create a design checkpoint (DCP).
7. That DCP is then submitted to AWS for generating an AWS FPGA image (AFI).
8. The AFI can then be simulated and debugged like any other executable script.
9. Amazon even allows the AFI to be shared with in its cloud regions.

Amazon F1 instances are reusable and sharable and can be deployed in a scalable manner to support a large model.

9.3.3 Dedicated AI Accelerators Using ASICs

While GPUs and FPGAs can perform better than CPUs for AI-related tasks, a further $10\times$ execution speed may be gained [4] with customized ASICs (application-specific integrated circuit). Such accelerators employ strategies to optimize memory access, and use lower precision arithmetic to speed up calculation and improve throughput. Examples of low-precision floating-point formats used for AI acceleration [15] include half-precision and bfloat to improve computational efficiency.

Traditional binary floating point format has a sign, a c, and an exponent. The sign bit indicates positive or negative values. A significand (whose fractional part is commonly known as the mantissa) is a binary fixed-point number of the form 0. abcd... or 1.abcd..., where the fractional part is represented by a fixed number of binary bits after the radix point. The exponent represents multiplication of the significand by a power of 2. AI systems are usually trained using 32-bit IEEE 754 binary32 single precision floating point [15], as shown in Fig. 9.5.

Reducing a 32-bit representation to 16 bits using half-precision (bfloat16) results in significant memory, performance, and energy savings.

Higher throughput translates to accelerated training. It also boosts productivity and saves energy. Community of hardware designers currently treat AI training and inference as two distinct tasks, each requiring two or more different architectures. Nvidia recently announced a new integrated architecture called Ampere A100 that supports multiple high-precision training formats, as well as lower precision formats commonly used for inference [16]. A comparison of A100 with previous generation architectures is shown in Table 9.1. Previously, V100 was announced in 2017 and P100 in 2016. As seen below, boost clock frequency went down in A100, but due to

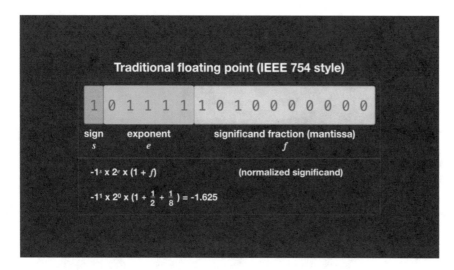

Fig. 9.5 Digital representation of a floating point number [15]

Table 9.1 Comparing three generations of Nvidia's AI hardware accelerators

	A100	V100	P100
FP32 CUDA cores	6912	5120	3584
Boost clock	~1.41 GHz	1530 MHz	1480 MHz
Memory clock	2.4 Gbps HBM2	1.75 Gbps HBM2	1.4 Gbps HBM2
Memory bus width	5120-bit	4096-bit	4096-bit
Memory bandwidth	1.6 TB/s	900 GB/s	720 GB/s
VRAM	40 GB	16 GB/32 GB	16 GB
Single precision	19.5 TFLOPS	15.7 TFLOPs	10.6 TFLOPs
Double precision	9.7 TFLOPs (1/2 FP32 rate)	7.8 TFLOPs (1/2 FP32 rate)	5.3 TFLOPs (1/2 FP32 rate)
INT8 tensor	624 TOPs	N/A	N/A
FP16 tensor	312 TFLOPs	125 TFLOPs	N/A
TF32 tensor	156 TFLOPs	N/A	N/A
Interconnect	NVLink 3 12 Links (600 GB/s)	NVLink 2 6 Links (300 GB/s)	NVLink 1 4 Links (160 GB/s)
GPU	A100 (826 mm^2)	GV100 (815 mm^2)	GP100 (610 mm^2)
Transistor count	54.2 B	21.1 B	15.3 B
TDP	400 W	300 W/350 W	300 W
Manufacturing process	TSMC 7 N	TSMC 12 nm FFN	TSMC 16 nm FinFET
Interface	SXM4	SXM2/SXM3	SXM
Architecture	Ampere	Volta	Pascal

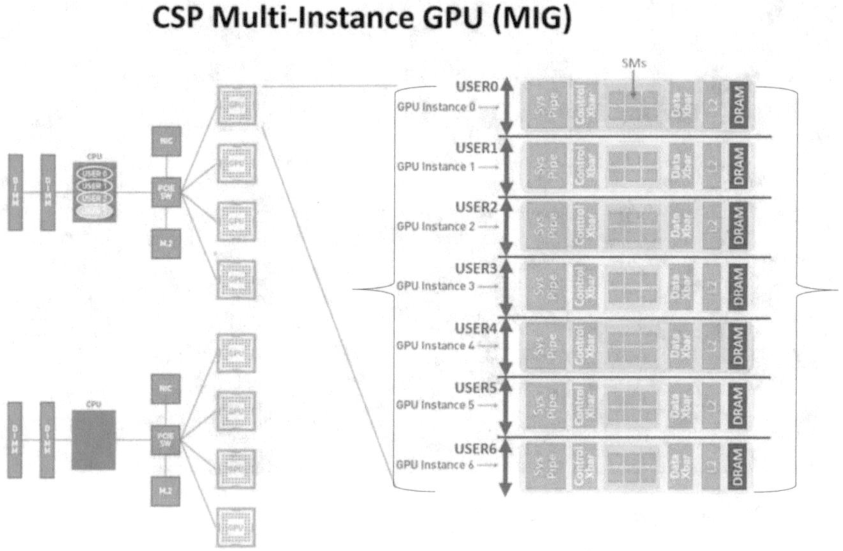

CSP Multi-Instance GPU (MIG)

Fig. 9.6 GPU logical partitioning using MIG

increased memory bus width, a higher bandwidth can be realized as compared to the previous two generations.

The leading A100 Ampere part is built using TSMC's 7 nm process technology with 54 billion transistors, 2.5× of what V100 had in the previous generation. While the operations in floating point format shows only a moderate improvement, the performance of tensor operations greatly improves by ~2.5× for FP16 tensors, as well as for the new 32-bit format called TF32. Memory speed got a significant expansion to deliver a total of 1.6 TB/s bandwidth. While this product consumes 400 W of power at full performance for AI training, it has a lower power mode for AI inference tasks. Furthermore, A100 can be scaled up with multiple accelerators using NVLink, or scaled out using NVIDIA's new Multi-Instance GPU (MIG) technology to split up a single A100 for multiple workloads.

MIG, as shown in Fig. 9.6, is a mechanism for GPU portioning. It enables a single A100 to be partitioned into up to seven virtual GPUs, each with a dedicated allocation of system memory, L2 cache, and memory controllers. This allows each user/task running in a partition its own set of dedicated resources with a predictable level of performance. This is a virtualization technology, which allows cloud service providers (CSPs) to allocate compute time on an A100 with full isolation between different tenants. It has a business implication of serving more users or applications on a single hardware without overprovision as a safety margin.

An example of different A100 instances that can be offered in a public cloud for different workloads is shown in Fig. 9.7. Highest requirement is for high-performance computing (HPC), which may require up to 40 GB of memory and only one task running exclusively on the A100. On the other end of a spectrum is

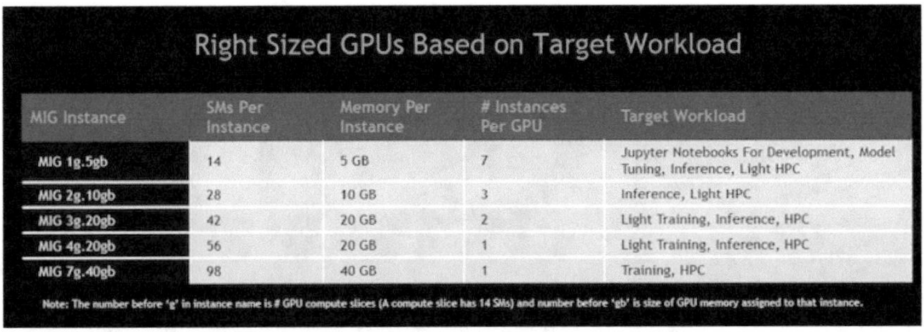

MIG Instance	SMs Per Instance	Memory Per Instance	# Instances Per GPU	Target Workload
MIG 1g.5gb	14	5 GB	7	Jupyter Notebooks For Development, Model Tuning, Inference, Light HPC
MIG 2g.10gb	28	10 GB	3	Inference, Light HPC
MIG 3g.20gb	42	20 GB	2	Light Training, Inference, HPC
MIG 4g.20gb	56	20 GB	1	Light Training, Inference, HPC
MIG 7g.40gb	98	40 GB	1	Training, HPC

Note: The number before 'g' in instance name is # GPU compute slices (A compute slice has 14 SMs) and number before 'gb' is size of GPU memory assigned to that instance.

Fig. 9.7 A spectrum of AI tasks running on an A100

Table 9.2 High-performance interlink technologies<<

	NVLink 3	NVLink 2	NVLink 1
Signaling rate	50 Gbps	25 Gbps	20 Gbps
Lanes/link	4	8	3
Bandwidth/direction/link	25 GB/s	25 GB/s	20 GB/s
Total bandwidth/link	50 GB/s	50 GB/s	40 GB/s
Links/chip	12 (A100)	6 (V100)	4 (P100)
Bandwidth/chip	600 GB/s	300 GB/s	160 GB/s

a Jupyter notebook, or a light-weight HPC task, or an inference job that needs only up to 5 GB of memory. Thus, up to seven such tasks can run simultaneously on an A100.

We previously mentioned NVLink as the interconnect technology for scaling up with multiple A100 accelerators. It is NVIDIA's proprietary high bandwidth solution that allows up to 16 GPUs to be connected for operation as a single cluster, for large workloads that need higher levels of performance. It is a third-generation technology, as shown in Table 9.2, offering up to 600 GB/s bandwidth/chip.

All of the above building blocks have an architecture, for example, eight GPU configuration, where each GPU is directly connected to every other GPU. This is shown in Fig. 9.8, using NVSwitches which support NVLink 3's faster signaling rates. According to NVIDIA, the first machine using this architecture has been delivered to Argonne National Laboratory. NVLink is a wire-based protocol serial multi-lane near-range communication link developed by Nvidia [17]. Using this, devices use mesh networking to communicate instead of a central hub.

A comparison of different HPC application speedups [18] as compared to NVIDIA's previous Tesla V100 is shown in Fig. 9.9.

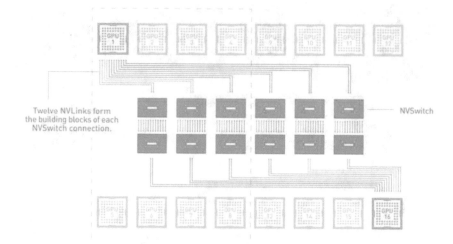

Fig. 9.8   A hybrid mesh cube design with A100 and fast interconnects

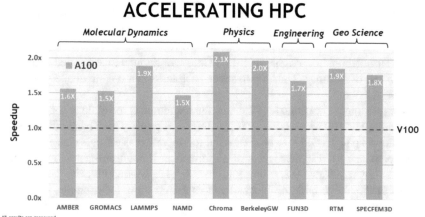

Fig. 9.9   A100 HPC application speedups as compared to previous generation

9.4   Cerebras's Wafer Scale AI Engine

Another notable effort to address areas requiring specialized HW for AI processing is by Cerebras [19]. They have chosen to tackle the scale-up issues by pioneering a single wafer scale engine (WSE), the largest chip ever built for deep learning systems. It is ~50× larger than other contemporary chips. The objective is to deliver more compute power, larger memory, and higher communication bandwidth [30].

Fig. 9.10 Comparison of Cerebras AI chip size with a GPU from Nvidia [19]

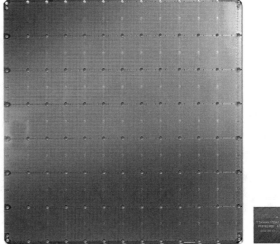

As shown in Fig. 9.10, their WSE measures 46,225 mm^2 with 1.2 trillion transistors and 400,000 AI-optimized cores. In comparison, the latest A100 GPU from Nvidia measures 826 mm^2 and has 54.2 billion transistors. This translates to 18 GB SRAM, 9.6 PB/s of memory bandwidth, and a System I/O capacity of 1.2 TB/s for Cerebras.

The WSE contains 400,000 sparse linear algebra compute (SLAC) cores. Each core is programmable and optimized for computations relevant for most neural networks. The Cerebras software platform integrates with ML frameworks such as TensorFlow and PyTorch. AI researchers and professionals may use a C++ interface to develop kernels to build custom neural networks. A Cerebras Graph Compiler (CGC) translates neural network to an optimized WSE executable. CGC does this by optimizing and mapping the given code to the WSE hardware. Each stage of CGC is designed to maximize WSE hardware utilization. Software kernels are used such that more compute resources are allocated in parallel to perform complex operations. CGC generates code placement for each neural network elements to minimize communication latency between layers. The layered architecture of the software platform is shown in Fig. 9.11.

As of this book's writing, two WSE systems (CS1s) were bought by Pittsburg Supercomputing Center [20]. In a datacenter, a rack unit (or 1 U) is the standard metric for servers and other networking equipment box sizes, such that 1 U = 1.75″ or 44.55 mm. Each WSE CS1 is a fully integrated 15 U chassis that requires 20 KW of power through 12 × 4 KW supplies with a built-in redundancy. These machines are expected to be installed in late 2020. This would enable AI researchers to train their models, covering domains such as healthcare for disease management and control, power generation, transportation, and many other socially relevant programs.

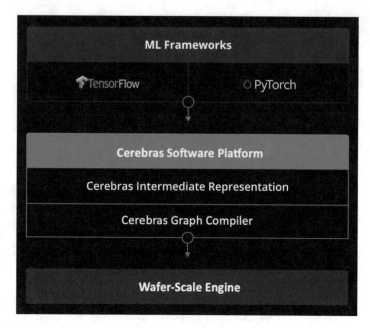

Fig. 9.11 Software stack for Cerebras WSE [19]

9.5 Google Cloud TPUs

Google has custom-designed machine learning tensor processing units (TPUs) [31] using application-specific integrated circuits (ASICs) that power its premium products [21] such as Translate, Photos, Search, Assistant, and Gmail. These machines are tailored for TensorFlow and have been used in Google datacenters since 2015. In comparison to a FPGA-based implementation, ASIC designs need high investment, but offer an order of magnitude better performance per watt for machine learning workloads. This represents about $10\times$ improvement over Moore's law, which translates to about 7 years of leap forward. As compared to a custom-designed CPU or GPU, a TPU has a reduced design cycle due to electronic design automation (EDA) tools. Specially, use of high-level synthesis and hardware design languages (HDL) such as Genesis II or Chisel helps with a rapid prototyping and system design exploration. A single ASIC in Google's TPU version3 is able to deliver 420 Teraflops of compute capacity using 128 GB of high-bandwidth memory (HBM). A single Cloud TPU Pod can include more than 1000 individual TPU chips which are connected by a two-dimensional mesh network, with a degree of 4. TPU nodes are laid out in a two-dimensional rectangular lattice of n rows and n columns, with each node connected to its four nearest neighbors, and the corresponding nodes on opposite edges are connected [22]. Thus, Google Cloud is able to deliver 100+ Petaflops, with 32 HBM on a 2D toroidal mesh network, shown in Fig. 9.12.

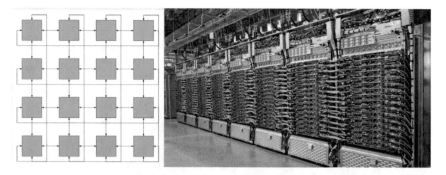

Fig. 9.12 Cloud TPUs laid out in a 2D mesh network (left) in a Pod (right) [21, 22]

Fig. 9.13 Amazon's
Inferentia Chip [24]

Neural network workloads must be able to run multiple iterations of the entire training loop on a TPU [32]. Although this is not a fundamental requirement of TPUs themselves, this is one of the current constraints of the TPU software ecosystem. Hopefully, this restriction will ease in the near future as software stacks evolve [33].

9.6 Amazon's Inference Engine

Following the general 80-20 rule, inference requires relatively a smaller fraction of compute as compared to training, but its share of infrastructure cost is >90%, while the rest 10% goes toward machine learning training infrastructure [23].

With this in mind, Amazon acquired Annapurna, an Israeli start-up in 2015. Engineers from Amazon and Annapurna Labs built the ARM Graviton processor and Amazon's Inferentia Chip shown in Fig. 9.13:

Inferentia chip consists of four Neuron Cores, each of which implements a systolic array matrix multiply engine. A systolic array is a homogeneous network of tightly coupled data processing units (DPUs) called cells or nodes. Each node or

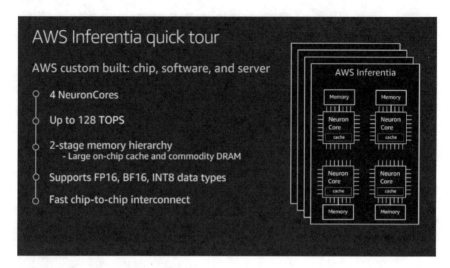

Fig. 9.14 Amazon's Inferentia Chip [24]

DPU independently computes a partial result as a function of the data received from its upstream neighbors, stores the results within itself and passes it downstream. A single Inferentia chip can perform up to 128 TOPS (trillions of operations per second). It supports BF16, INT8, and FP16 data types. This chip can take a 32-bit trained model and run it at the speed of a 16-bit model using BFloat 16 [24], as depicted in Fig. 9.14. As ML model sizes grow, transferring a model in and out of memory becomes crucial due to latency issues. This is solved by using chip interconnect and partitioning a model, then mapping it across multiple cores with 100% on-cache memory usage. It allows data to stream at full speed through the pipelines of cores avoiding the latency issues caused by external memory accesses.

Inferentia is being used in Amazon's AWS Cloud using a variety of frameworks. The model needs to be compiled to a hardware-optimized representation. Operations can be performed through command-line tools available in the AWS Neuron SDK or via framework APIs.

Inferentia uses ASIC design technology, similar to Google's TPUs, giving it an order of magnitude latency and power/performance advantage over FPGA or general-purpose CPU-based ML solutions.

9.7 Intel's Movidius VPU

Computer vision is critical for many smart connected devices. Intel has acquired Movidius with vision processing units (VPUs), with its architecture shown in Fig. 9.15.

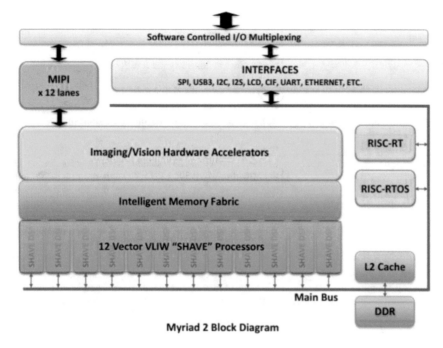

Fig. 9.15 Movidius VPU architecture [25]

Movidius VPUs enable edge AI vision processing using workload-specific hardware acceleration. It can process up to one trillion deep neural network (DNN) operations/second. Its architecture is designed to minimize data movement, to achieve a balance of power consumption and compute performance. This is achieved with SHAVE (streaming hybrid architecture vector engine) cores within the SoC. These support very long instruction (VLIW) words, wherein a compiler packs multiple sequential statements which can all be executed in parallel with no data dependencies. This helps parallel hardware units to execute the VLIW group of instructions in a single step, thereby achieving a higher throughput and performance. Readers may note that VLIW often leads to multiple instructions, multiple data (MIMD) mode of parallel execution. Latest product in the Movidius family is Myriad X, operating within a 2 W power envelope. This is much less than the power consumed by the previously discussed products, such as Nvidia's GPUs or Google's TPUs. However, the focus of Movidius products is not on the training but inference. Training workloads will require the processing power of a GPU or TPU. Salient features of Movidius Myriad X are shown in Table 9.3.

An interesting use-case that Intel has reported with Movidius is to catch poachers in Africa [27]. Detection cameras are battery powered and installed at various locations in jungle. A camera wakes up when it detects motion, and using an on-device AI algorithm is able to analyze images in real time. Then it alerts park headquarters when humans or vehicle are identified in any of the captured frames.

Table 9.3 Architectural features of Intel's Movidius Myriad X Chip [26]

Features	Benefits
Neural Compute Engine	With this dedicated on-chip accelerator for deep neural networks, the Movidius Myriad X VPU delivers over 1 trillion operations per second of DNN inferencing performance. Run deep neural networks in real time at the edge without compromising on power consumption or accuracy
16 Programmable 128-bit VLIW Vector Processors	Run multiple concurrent imaging and vision application pipelines with the flexibility of 16 vector processors optimized for computer vision workloads
16 Configurable MIPI Lanes	Connect up to eight HD resolution RGB cameras directly to the Movidius Myriad X VPU with support for up to 700 million pixels per second of image signal processing throughput
Enhanced Vision Accelerators	Utilize over 20 hardware accelerators to perform tasks such as optical flow and stereo depth without introducing additional compute overhead. For example, the new stereo depth accelerator can simultaneously process six camera inputs (three stereo pairs) each running 720p resolution at 60 Hz frame rate
2.5 MB of Homogenous On-Chip Memory	The centralized on-chip memory architecture allows for up to 400 GB/s of internal bandwidth, minimizing latency and reducing power consumption by minimizing off-chip data transfer

The use of AI and real-time inference provides a better deterrent than the currently prevalent method of screening or banning export/import of products made from already-dead animals. This newer approach prevents the animals' death in the first place.

9.8 Apple's AI Ecosystem

Apple has acquired over 20 AI companies since 2010, more than Google, Microsoft, Facebook, and Amazon [28]. Even though Apple's AI applications are less visible than these other companies, as its main focus has been on improving iPhone with new features. Siri is an example, as it was acquired as a virtual assistant of IOS operating system, for voice-based user support. Since then, the launch of FaceID has enabled face-tracking features in its latest consumer products.

Apple's AI focus is more on the edge-based and wearable consumer devices, such as detection of a potential heart attack for its watch users. One of its latest acquisition is Xnor.ai, whose edge AI engine previously enabled Wyze camera. It can recognize people, pets, and objects. However, unlike Amazon's Ring or Google's Nest Cam, the Wyze processes the images locally inside the camera for better privacy and security.

Unfortunately, due to Apple's insistence on secrecy, not much is known or published externally about its AI hardware features. However, their latest iPhone uses A13 processor [29]. Its SoC boosts performance by 20% and lowers power up to 40% over the previous year's phone models. However, most significant is the new Deep Fusion technology. It is Apple's version of neural image processing. It works by shooting multiple frames from new phone's three cameras and then combines them into a single picture. This helps to negate image noise, e.g., multi-colored dots that can appear in images. As noise does not appear identically in each frame, the system is able to select the least noise-ridden parts into the composed image for a cleaner and sharper result. It uses the built-in neural engine to enhance texture, scrub out blurriness, and sharpen details [29]. As a result, Apple uses processing, rather than packing more pixels into a sensor, to produce the best results [34].

9.9 Summary

AI and ML implementations are limited by lack of data, high model complexity, and available compute resources. The last two factors can be alleviated by using hardware accelerators. These can be based on GPUs, FPGAs, or ASIC-based designs. In future, the need for AI training and inference can be combined in a single piece of hardware. This will reduce both the capital expenditure (CAPEX) and operational expenditure (OPEX) in a data-center. To summarize, for training, CPU-based cloud operations are good; for classification and prediction on real-time data, GPUs are good; and for persistent data (like image archives), TPUs are good.

9.10 Points to Ponder

1. Can faster AI and ML solutions be used to make up for paucity of data?
2. Why AI training computation needs are higher than for inference?
3. How can an FPGA-based AI and ML solution outperform GPA?
4. Why does an ASIC solution offer higher performance than FPGA?
5. How a high bandwidth memory may reduce the need for communication between chips?

References

1. https://ieeexplore.ieee.org/popular/all.
2. https://en.wikipedia.org/wiki/AI_winter.
3. https://www.bloomberg.com/professional/blog/rise-of-artificial-intelligence-and-machine-learning/.
4. https://en.wikipedia.org/wiki/AI_accelerator.
5. Mittal, S. (2018). A survey of ReRAM-based architectures for processing-in-memory and neural networks. *Machine Learning and Knowledge Extraction, 1*(1), 75–114.

6. Sze, V., Chen, Y.-H., Emer, J., Suleiman, A., & Zhang, Z. (2017). Hardware for machine learning: Challenges and opportunities, October 2017. https://arxiv.org/pdf/1612.07625.pdf.
7. https://www.design-reuse.com/articles/46634/re-architecting-socs-for-the-ai-era.html.
8. https://www.computer.org/publications/tech-news/chasing-pixels/is-it-time-to-rename-the-gpu.
9. https://en.wikipedia.org/wiki/Graphics_processing_unit.
10. https://en.wikipedia.org/wiki/SIMD.
11. Taylor-Weiner, A., Aguet, F., Haradhvala, N. J., Gosai, S., Anand, S., Kim, J., et al. (2019). Scaling computational genomics to millions of individuals with GPUs. *Genome Biology, 20*, 228. https://doi.org/10.1186/s13059-019-1836-7.
12. Mittal, S., & Vaishay, S. (2019). A survey of techniques for optimizing deep learning on GPUs. *Journal of Systems Architecture*. https://www.academia.edu/40135801/A_Survey_of_Tech niques_for_Optimizing_Deep_Learning_on_GPUs.
13. Chung, E., Strauss, K., Fowers, J., Kim, J.-Y., Ruwase, O., Ovtcharov, K. (2015, February 23). Accelerating deep convolutional neural networks using specialized hardware.
14. Official repository of the AWS EC2 FPGA Hardware and Software Development Kit, https://github.com/aws/aws-fpga.
15. https://engineering.fb.com/ai-research/floating-point-math/.
16. https://www.anandtech.com/show/15801/nvidia-announces-ampere-architecture-and-a100-products.
17. https://en.wikipedia.org/wiki/NVLink.
18. https://www.nvidia.com/content/dam/en-zz/Solutions/Data-Center/nvidia-ampere-architecture-whitepaper.pdf.
19. https://www.cerebras.net/.
20. https://www.anandtech.com/show/15838/cerebras-wafer-scale-engine-scores-a-sale-5m-buys-two-for-the-pittsburgh-supercomputing-center.
21. https://cloud.google.com/tpu.
22. https://en.wikipedia.org/wiki/Torus_interconnect.
23. https://www.cloudmanagementinsider.com/amazon-inferentia-for-machine-learning-and-artifi cial-intelligence/.
24. https://perspectives.mvdirona.com/2018/11/aws-inferentia-machine-learning-processor/.
25. https://www.extremetech.com/computing/254772-new-movidius-myriad-x-vpu-packs-custom-neural-compute-engine.
26. https://www.intel.com/content/www/us/en/products/docs/processors/movidius-vpu/myriad-x-product-brief.html.
27. https://www.intel.com/content/www/us/en/artificial-intelligence/solutions/fighting-illegal-poaching-with-purpose-built-ai-camera.html.
28. https://www.fool.com/investing/2020/01/23/apple-is-quietly-expanding-its-artificial-intellig. aspx.
29. https://www.electronicdesign.com/technologies/embedded-revolution/article/21808543/apples-a13-processor-powers-its-latest-iphone-lineup.
30. https://www.darpa.mil/our-research.
31. https://cloud.google.com/tpu/docs/tpus.
32. https://www.tensorflow.org/tutorials/customization/basics.
33. https://medium.com/sciforce/understanding-tensor-processing-units-10ff41f50e78.
34. https://www.pocket-lint.com/phones/news/apple/149594-what-is-apple-deep-fusion.

Appendix A

AI/ML for App Store Predictions

Using Python for App Metrics Predictions in Google Play Store[1]

Background: Programmers invest significant financial and human capital into maintaining their mobile applications (aka App). An app is often the only online touchpoint a business customer has with the company sponsoring that app, e.g., bank transactions. To be able to predict app store performance enables businesses to best prioritize resources and maximize their return on investment ("ROI"). We believe that leveraging app store data will help to predict critical app store performance metrics. Specifically, we will explore machine learning models to try and predict:

- *Downloads*: How many mobile app downloads can be expected?
- *Category*: Can we predict the classification category of the app?
- *Rating*: Are we able to predict how the app will be rated by customers?

Procedure: To complete our analysis we followed the standard Data Science Lifecycle [1]. This process includes five major tasks:

1. Obtain data
2. Scrub and prepare the data
3. Explore data
4. Model data
5. Interpreting data

---

[1]*Acknowledgment*: This work is based on a class project done by Andrew Herman and Adrian Berg in a class taught by our lead author, Prof. Pramod Gupta, at UC Berkeley in July 2019.

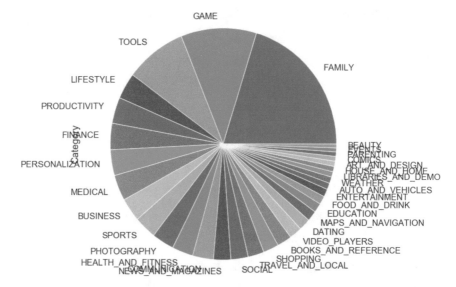

Fig. A.1 Distribution of Google Play Store applications by categories [2]

Each of these tasks will be explored in greater detail throughout this Appendix. It is important to note that the subsequent data manipulation and analyses were completed using Python 3.6, along with standard Python data science packages (e.g., Numpy, Pandas, Sklearn, etc.).

Step 1: Obtain Data

Obtaining data is straightforward, as it comes from a publicly available Kaggle dataset called "Google Play Store Apps" uploaded by Lavanya Gupta [2]. The datasets were captured by scraping publicly available data from the Google Play Store, as depicted in Fig. A.1. There are two distinct datasets that were used in these analyses:

1. Play Store App Data (10,840 records).
2. App User Review Data (64,300 records).

When conducting data analysis, it is important to understand what each column of data represents. The Kaggle website included descriptions for every column across both datasets.

Step 2: Scrub and Prepare the Data

Scrubbing and preparing the data can be a single, most time-consuming task. It is also the most important step. Without having clean data to input into our models,

Table A.1 Google Play Store data prep tasks

Cleanup task	Application data	User review data
Null value removal/fills	✓	✓
Duplicate removal	✓	✗
Data type conversions	✓	✓
Dummy variables	✓	✗
Rescale/normalize	✓	✗

results will be adversely impacted by outliers, introducing errors. The tasks required to be completed for each dataset are shown below in Table A.1.

Following adjustments are needed in this dataset as part of the data preparation:

1. Removal of 1488 nulls and empty cells.
2. Removal of 1170 duplicate records that were either:

   (a) Exact duplicates (876)
   (b) Slight variations (294)

3. Altering the data type, as only 1 out of the 13 features were of the correct data type.
4. Creating two separate datasets to run through our models, one with all of the application data and one with applications and associated user review data.

   Greater details on each of these tasks are as follows.

Step 2.1: Null Value Removal/Fills

The Play Store App Data did not have many null values. Majority of the null values were located within the rating column, representing 13.6% of the total database, as shown in Fig. A.2.

Unfortunately, because the rating column is a continuous variable, we had to remove these observations from the dataset, rather than assigning an arbitrary value (e.g., 999) to represent the nulls as a category. *Dropping these observations allowed us to eliminate the disruption they may have created within the model.*

Step 2.2: Duplicate Removal

We observed two types of duplicated observations, each requiring the removal of the duplicated entry from the dataset. In total, we observed that there were 8190 unique apps compared to 9360 total apps, which indicates 1170 duplicates in the dataset, as shown in Fig. A.3.

```
1  #First we can check out where some of the nulls are
2  df_null = df.isnull()
3  df_null.sum()
```

```
1  #We can sort and aggregate our nulls to find out how many and
2  #what percentage of our data they represent.
3  tot = df.isnull().sum().sort_values(ascending=False)
4  per = (df.isnull().sum()/df.isnull().count()).sort_values(ascending=False)
5  miss = pd.concat([tot, per], axis=1, keys=["Total","Percent"])
6  print(miss)
```

```
1  #We can plot this to see where our missing values are
2  f, ax = plt.subplots(figsize=(15,6))
3  plt.xticks(rotation="45")
4  sns.barplot(x=miss.index, y=miss["Percent"], palette="Blues_d")
5  plt.xlabel("Features", fontsize=14)
6  plt.ylabel("Percent of missing values", fontsize=14)
7  plt.title("Percent Missing Data for Each Feature", fontsize=16)
```

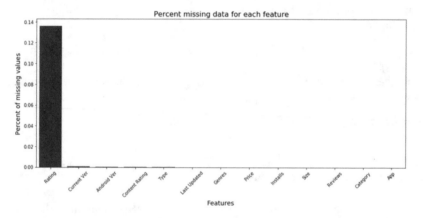

Fig. A.2 Steps to remove null values and the results

```
1  #We can start by comparing the number of unique apps to the total number of apps
2  unique = len(df_clean["App"].unique().tolist())
3  print("Nuber of unique apps:", unique)
4  print("Number of total apps:", df_clean.shape[0])
5  difference = df_clean.shape[0] - unique
6  print("So we expect", difference, "duplicates in this dataset")
```

```
1  #If we want to look at which rows are actual duplicates
2  exact_duplicates = df_clean[df_clean.duplicated(keep=False)]
3  exact_duplicates = exact_duplicates.sort_values(by=["App"], axis=0, ascending=True)
4  print(exact_duplicates.head(6))
```

```
1  example = df_clean.loc[df_clean["App"] == "ESPN"]
2  print(example)
```

Fig. A.3 Column-wise data cleaning and preparation

Exact Duplicates (876)

These are apps where the entire observation is identical. We are able to easily filter this by seeing apps with the exact same name and number of reviews.

	App	Category	Rating	Reviews	Size
1407	10 Best Foods for You	HEALTH_AND_FITNESS	4.0	2490	3.8M
1393	10 Best Foods for You	HEALTH_AND_FITNESS	4.0	2490	3.8M
2543	1800 Contacts – Lens Store	MEDICAL	4.7	23160	26M
2322	1800 Contacts – Lens Store	MEDICAL	4.7	23160	26M
2256	2017 EMRA Antibiotic Guide	MEDICAL	4.4	12	3.8M
2385	2017 EMRA Antibiotic Guide	MEDICAL	4.4	12	3.8M

Slight Variations (294)

These are apps where the name and all other details are identical, except for the number of reviews. To eliminate the duplicates, we first sorted by the highest number of reviews and then removed duplicates by app name. This kept the record with the highest number of reviews, which we believe to be accurate.

	App	Category	Rating	Reviews	Size	Installs	Type
2959	ESPN	SPORTS	4.2	521138	Varies with device	10,000,000+	Free
3010	ESPN	SPORTS	4.2	521138	Varies with device	10,000,000+	Free
3018	ESPN	SPORTS	4.2	521138	Varies with device	10,000,000+	Free
3048	ESPN	SPORTS	4.2	521140	Varies with device	10,000,000+	Free
3060	ESPN	SPORTS	4.2	521140	Varies with device	10,000,000+	Free
3072	ESPN	SPORTS	4.2	521140	Varies with device	10,000,000+	Free
4069	ESPN	SPORTS	4.2	521081	Varies with device	10,000,000+	Free

Step 2.3: Data Type Conversions

We converted the majority of the fields into continuous values to be used in the algorithms. In the literature, application ratings can be treated as a continuous or ordinal categorical variable, and both methods were tested to observe which approach offers better insights.

Some required manipulation to convert stored values into number (e.g., $1.00 is stored as a string), while others required numerical dummy categories (e.g., genres).

Step 2.4: Dummy Variables

As previously stated, the majority of the data types were objects / strings. Some features such as app category were converted into numerical values to work within the models. To do this, we used the LabelEncoder function. For conversion of the "Sentiment" feature, we created custom categories to represent the ordinal nature of the data.

Step 2.5: Rescale

As a result of the features with highly variable units, magnitudes, and range, there are tradeoffs inherent in the method used to scale. For example, many of the ML algorithms such as KNN rely on Euclidean distance for computations, which can be significantly impacted by the magnitude. To account for this influence, we normalized each feature to prepare the data to suit the algorithms. We elected to use Min–Max scaling for multiple features, as this method can work well for non-normally distributed data, or when the standard deviation is relatively small. Its major limitation is that it suppresses outliers; however, this feature did not have any outliers needing removal. The chart given displays the distribution of applications by size, as computed and shown in Fig. A.4.

As the app size data is not normally distributed, we must do so by using the MinMax function, as shown in Fig. A.5:

Step 3: Data Exploration

Data exploration is a critical step within the Data Science Lifecycle. This allows us to not only understand the data and the relationships between features, but also to spot outliers and other suspicious looking patterns that may indicate an abnormality. Some high-level "fun facts" about the data are:

1. Unique apps: 9659
2. Unique categories: 32
3. Unique genres: 48
4. Median number of reviews per app: 2094

   (a) The mean is 444k, which indicates significant outliers.

5. Average app price: $1.027
6. Apps with a perfect 5.0 rating: 274

Additionally, we looked holistically at the correlations of the data, as shown in Fig. A.6. It is concerning that most of the features have low correlation, except for Rating and Reviews.

```
1  temp_df = pd.DataFrame(df_cleaner['Size'])
2  temp_df['SizeTrim'] = [x[:-1] for x in temp_df['Size']] #remove last character, new col
3  temp_df['SizeUnit'] = [x[-1] for x in temp_df['Size']] #Separate out last character, new col
```

```
1  print(temp_df['SizeUnit'].unique())
2  temp_df.loc[temp_df.SizeUnit == 'M', 'multiplier'] = 1000 #replace values
3  temp_df.loc[temp_df.SizeUnit == 'k', 'multiplier'] = 1 #replace values
4  temp_df["multiplier"] = temp_df["multiplier"].fillna(0) #replace values
5  temp_df["multiplier"]= temp_df["multiplier"].astype(int) #Change data type
```

```
1  temp_df.loc[temp_df.multiplier == 0, 'SizeTrim'] = np.NaN #replace values
2  temp_df["SizeTrim"]= temp_df["SizeTrim"].astype(float)
3  temp_df["NewSize"] = temp_df["SizeTrim"] * temp_df["multiplier"] #calc new size
```

```
1  print(temp_df.dtypes)
2
3  size_median = round(temp_df["NewSize"].median(skipna = True),0) #calc median
4  temp_df["NewSize"] = temp_df["NewSize"].fillna(size_median) #replace missing
5
6  df_cleaner['Size'] = temp_df["NewSize"].astype(int) #update original data
```

```
1  plt.hist(df_cleaner["Size"], bins=30, log=True, color="b")
2  plt.xlabel("Size")
3  plt.ylabel("# of Applications With This Size")
4  plt.xlim((0,100000))
5  plt.title("Distribution of Applications by Size")
```

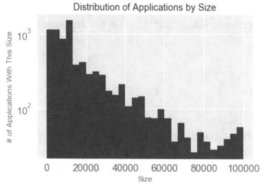

Fig. A.4 Google Apps distribution by their binary sizes

Step 4: Categorization

The chart in Fig. A.7 illustrates the distribution of apps by category. Some interesting observations are as follows:

1. It is surprising that family, the largest genre, is nearly as large as games (2) and tools (3) combined. We considered the potential for feature imbalance; however, relative to the other categories, it does not appear to be a large proportion of records.
2. Some categories could theoretically be merged (e.g., family + parenting, travel and local + maps and navigation). Having multiple categories that essentially mean the same thing will make the model less accurate. Unfortunately, we do not have definitions available of each category, so we are unable to confidently make the assumption that it would be safe to merge these. We will have to proceed with

```
1  plt.hist(df_cleaner["Size_s"], bins=30, log=True, color="b")
2  plt.xlabel("Size_s")
3  plt.ylabel("# of Applications With This Size")
4  plt.xlim((0,1))
5  plt.title("Distribution of Applications by Size")
```

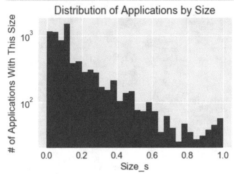

Fig. A.5 Normalized Google Apps distribution by their binary sizes

this analysis knowing there is a chance that multiple names for the same category are being used.

Step 4.1: Google App Ratings

App ratings are skewed lower; however, there is a small bump on the negative rating. This is in-line with the understanding that people tend to bias ratings higher [4]. The data used in the chart on the right of Fig. A.8 has been scaled using a MinMaxScaler to fit into the desired (0,1) range.

Furthermore, looking at Sentiment and Sentiment Polarity from the App User Review Data, as shown in Fig. A.9, also supports our hypothesis that people bias ratings are higher [4].

Step 4.2: Google App Reviews

Reviews have a long tail, as shown in Fig. A.10. It means that most apps have a low number of reviews, while a select few have an extremely high number of reviews. The large range of observations within this feature will have an adverse impact on the model's performance, so a z-score-based removal of outliers was also performed prior to rescaling.

```
1  corrmat = df_subset_s[["Installs_s","Rating_s","Reviews_s","Size_s","Price_s","Android Ver",
2                         "avg_polarity", "avg_sent_polarity"]].corr()
3  f, ax = plt.subplots(figsize = (10,10))
4  sns.heatmap(corrmat, annot=True, ax=ax, cmap="YlGnBu")
5  plt.title("Heatmap for Subset Data")
6  plt.savefig("Heat Map.png")
```

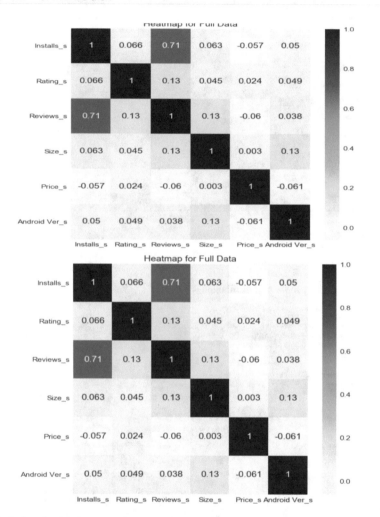

Fig. A.6 Google Play Store App Dataset details

Step 4.3: Google App Sizes

The distribution of apps by size is as expected and shown in Fig. A.11. There are a
high number of "small" apps, and the count of apps decreases as size increases. We
observed that on the far right the counts of large apps begin to increase. We are
unsure why this behavior occurs, but we do not believe these apps are outliers.

```
1  #We can plot this to see where our missing values are
2  category_size.reset_index(drop=False)
3  f, ax = plt.subplots(figsize=(15,6))
4  plt.xticks(rotation="45", ha="right")
5  sns.barplot(x=category_size.index, y=category_size, palette="Blues_d")
6  plt.xlabel("Category", fontsize=14)
7  plt.ylabel("Number of apps in category", fontsize=14)
8  plt.title("Distribution of Apps by Category", fontsize=16)
```

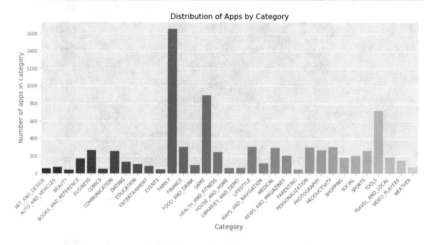

Fig. A.7 Google Play Store App distribution

```
1  ax = sns.distplot(df_cleaner["Rating"], axlabel="App Rating", bins=30, color="b")
2  ax.set_xlabel('App Rating')
3  ax.set_ylabel('Log Frequency')
4  ax.set_title("App Rating and Frequency")
```

```
1  from sklearn import preprocessing
2  min_max_scaler = preprocessing.MinMaxScaler()
3  df_cleaner["Rating_s"] = min_max_scaler.fit_transform(df_cleaner[["Rating"]])
```

```
1  ax = sns.distplot(df_cleaner["Rating_s"], axlabel="App Rating", bins=30, color="b")
2  ax.set_xlabel('Scaled App Rating')
3  ax.set_ylabel('Log Frequency')
4  ax.set_title("Scaled App Rating and Frequency")
```

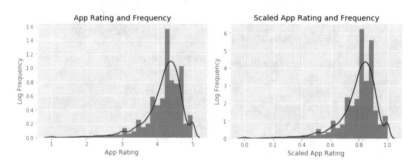

Fig. A.8 Google App Ratings distribution

```
1  plt.figure(figsize=(8,6))
2  plt.hist(user_reviews['Sentiment_s'], color="blue", align="left")
3  plt.show()
```

```
1  plt.figure(figsize=(8,6))
2  plt.hist(user_reviews['Sentiment_Polarity'], color="blue")
3  plt.show()
```

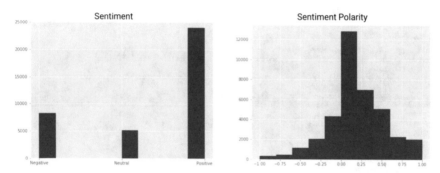

Fig. A.9 Google App Rating sentiments

```
1  df_cleaner = df_cleaner[(stats.zscore(df_cleaner["Reviews"]) < 3)]
```

Check the shape (previously had 8190 rows):

```
1  df_cleaner.shape
```

`(8133, 15)`

```
1  plt.hist(df_cleaner["Reviews"], bins=30, log=False, color="b")
2  plt.xlabel("Reviews")
3  plt.ylabel("Number of Apps")
4  plt.xlim(0,2000000)
5  plt.title("Histogram of Reviews With Anomalies Removed")
```

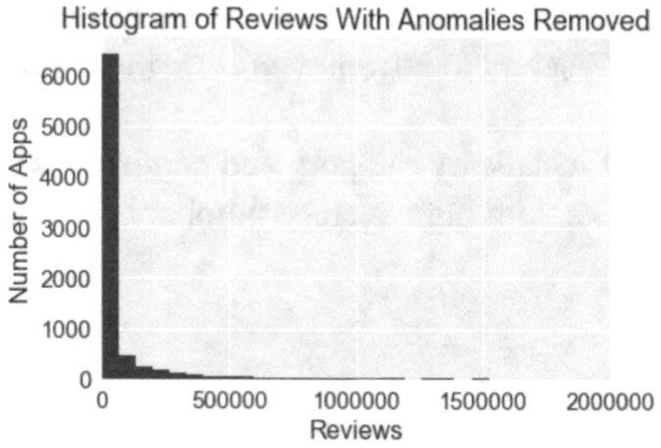

Fig. A.10 Google App Reviews distribution

```
1  print(temp_df.dtypes)
2
3  size_median = round(temp_df["NewSize"].median(skipna = True),0) #calc median
4  temp_df["NewSize"] = temp_df["NewSize"].fillna(size_median) #replace missing
5
6  df_cleaner['Size'] = temp_df["NewSize"].astype(int) #update original data
```

```
Size             object
SizeTrim         float64
SizeUnit         object
multiplier       int64
NewSize          float64
dtype: object
```

```
1  plt.hist(df_cleaner["Size"], bins=30, log=True, color="b")
2  plt.xlabel("Size")
3  plt.ylabel("# of Applications With This Size")
4  plt.xlim((0,100000))
5  plt.title("Distribution of Applications by Size")
```

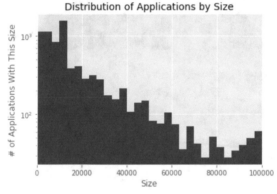

Fig. A.11 Google App distribution by size in bytes

Step 4. 4: Google App Installs

Installs behavior is as expected; however, we did notice a small number of outliers several standard deviations above the mean that had to be removed from the dataset, as shown in Fig. A.12.

We also looked at the average number of installs by category and confirmed that nothing looked suspicious (e.g., the categories with high average number of installs make sense), as shown in Fig. A.13.

Step 4.5: Type

This was a binary datatype comprised of "Free" or "Paid." We removed this entire feature from the dataset because it was redundant to the Price feature (i.e., $0.00 = $ "Free" and $>$0.00 $=$ "Paid").

```
1  plt.scatter(df_cleaner["Installs"],df_cleaner["Rating"], color="b")
2  plt.xlabel("Installs")
3  plt.ylabel("Rating")
4  plt.title("Installs by Rating")
```

```
1  plt.scatter(df_cleaner["Installs"],df_cleaner["Rating"], color="b")
2  plt.xlabel("Installs")
3  plt.ylabel("Rating")
4  plt.title("Installs by Rating With Outliers Removed")
```

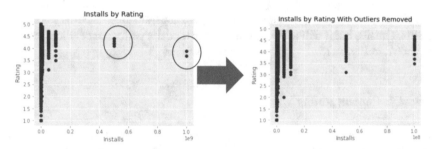

Fig. A.12 Google App installs by size

```
1  #Let's take a look at the distribution of our apps across categories
2
3  plt.figure(figsize=(8,6))
4  df_cleaner.groupby("Category")["Installs"].mean().plot(kind="bar", color="b")
5  plt.title("Installs by Category")
6  plt.ylabel("Average number of Installs")
7
8  plt.show()
9
```

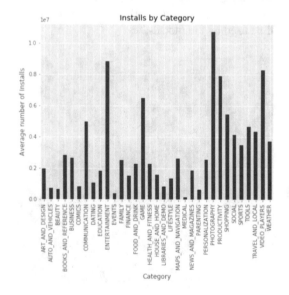

25

Fig. A.13 Google App installs by categories

Step 4.6: Price

Price had a small number of extreme outliers, as shown in Fig. A.14. These are required to be removed. The relatively small number of records with a price also created significant restrictions on using this feature as a target in any model, and as a result, this line of inquiry was not pursued.

The distribution of prices for apps was also far normal, with only 7.35% of the apps being paid, and the majority of those priced at $0.99. This makes sense from a practical perspective (e.g., many of the largest apps are free, such as Facebook, ESPN, Gmail), but it will adversely impact the model's performance. Final result is as shown in Fig. A.15.

Step 4.7: Content Rating

This was a categorical feature which we converted to ordinal, allowing us to potentially do some future analysis with this data. The gradation between a rating for "Everyone" and "PG-13" suggests an inherent order to the categories that we sought to capture in our analysis.

Step 4.8: Genres

The dataset contained numerous Genre observations that included two values. We split these apart into "Genre1" (47 unique values) and "Genre2" (6 unique values). Lastly, we converted these observations into numerical values using the LabelEncoder, so that they could be used by the models.

```
1  plt.figure(figsize=(8,6))
2  plt.scatter(df_cleaner["Rating"], df_cleaner["Price"], color="b")
3  plt.xlabel("Rating")
4  plt.ylabel("Price ($)")
5  plt.title("Rating and Price")
6  plt.show()
```

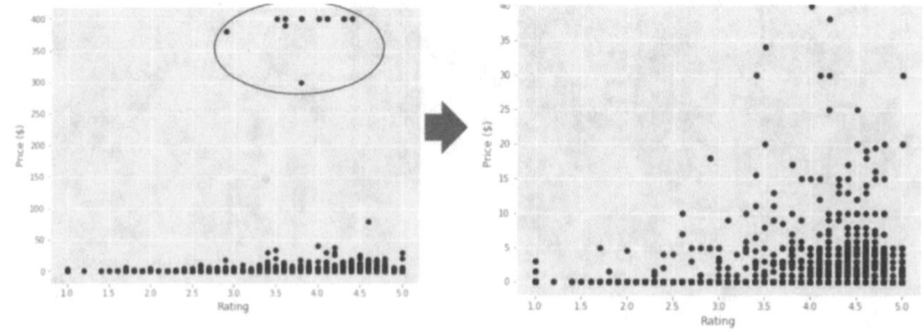

Fig. A.14 Google App Price Outlier removal

```
1  nonzero = df_cleaner[df_cleaner["Price"] !=0]
2  prices = nonzero.Price
3  plt.figure(figsize=(8,6))
4  plt.hist(prices, color="blue", bins = 20)
5  plt.xlim(0,400)
6  plt.title("Prices for Paid Apps")
7  plt.xlabel("Price in $")
8  plt.ylabel("Number of Apps")
```

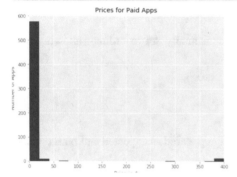

Fig. A.15 Google App installs by prices

Step 4.9: Last Updated

This feature was simply converted into a datetime format for usability within the models.

Step 4.10: Current Ver

We did not consider this attribute.

Step 4.11: Android Ver

We trimmed the original string of the Android Version down to one decimal. For example, Android Version 5.4.7.2 became Android Version 5.4. This allowed us to easily convert the datatype to a numerical value. We are confident this would not impact app store performance, as the variations of Android that we grouped together (up to one decimal place) are typically bug and/or security fixes, rather than a major feature update impacting customer usage.

App User Review Dataset Details:

This dataset contained 26,863 null values which we had to drop. This accounts for approximately 42% of the 64,295 total rows of data.

Step 4.12: Translated Review

The user reviews are not helpful for the model so we removed them entirely from the dataset.

Step 4.13: Sentiment

This was a categorical feature which we manually set to ordinal, allowing us to incorporate into the models.

Step 4.14: Sentiment Polarity

As stated previously, the sentiment polarity appeared as expected and biased toward positive reviews.

Step 4.15: Sentiment Subjectivity

This feature is extremely similar to sentiment polarity, so we removed it from the dataset as we felt it was a redundant data field.

Step 4.16: Merging the Data

The find step was to merge the two datasets into one. We used an "Outer Join" to merge on the "App" feature as the key.

Step 5: Results

Each of the areas of inquiry we sought to pursue were yielded underwhelming results, and no reliable predictive model was able to be developed.

Step 5.1: Predicting Installs

For this model, we decided to use a multiple linear regression (MLR) to predict the number of installs. This approach allows us to predict the number of app installations as a continuous variable. A preliminary look at pairwise Pearson's R for this dataset indicate that only polarity and sentiment polarity have any great issues with multicollinearity; for this reason, sentiment polarity is left out.

An initial OLS regression was run for the full dataset (excluding user sentiment data), and the subset data (inclusive of the user sentiment data) to observe which performed better, and assess p-values for further model refinement. The full data yielded an R-squared of 0.621, and our subset data yielded an R-squared of 0.666. After re-running the model based on the full dataset, the resulting R-squared is 0.621 and adjusted R-squared of 0.620, suggesting the model is a poor predictor of the number of installations. It is a good practice to check the normality of errors of a linear regression after modeling to ensure that errors follow a normal distribution with a mean of zero. However, due to the poor performance of this model, this validation step is unnecessary.

Step 5.2: Predict Category

The next model we tried to predict is the App Category. For this type of analysis, we sought to apply several different classifiers to observe which would work best. To start, we dropped the data into a Random Forest Classifier which, as an ensemble method, can provide a general sense of the comparative accuracy from using the complete data (accuracy score: 0.200) or the subset dataset (accuracy score: 0.120). Leveraging the accuracy score of each model as a rough measure using the holdout method (80%/20% split), we decided to proceed with the larger dataset and run several different models to try to improve our ability to predict the category.

We elected to run Logistic Regression ("LR"), Random Forest ("RF"), K-Nearest Neighbors ("KNN"), DT Classifier ("DT"), and Support Vector Machines ("SVM"). Due to some of these algorithms being affected by varying scales and magnitudes, the rescaled data was utilized for all of the models to allow for more side-by-side comparisons. Gaussian Naïve Bayes was not run due to the heavier reliance on data fitting a normal distribution. The comparative analysis was also run using cross-validation to help reduce the over-fitting of each model. Running these models proved generally underwhelming, with Logistic Regression yielding the highest accuracy score at 0.221, as shown in Fig. A.16.

Using Testing Accuracy as another performance metric, we find that Logistic Regression is still the best performer with an accuracy score of 0.205. The mean log loss of -3.038 (STD 0.019) further confirms the unreliability of the model.

Step 5.3: Predict Rating

The last feature we attempted to explore was predicting the application's rating. For this analysis, two approaches are supported by the literature:

1. Treat app ratings as a continuous variable
2. Treat ratings as ordinal data

First, we will run full dataset with Random Forests to determine the importance of different features of an app, as shown in Fig. A.17.

```
1  # boxplot algorithm comparison
2  fig = plt.figure()
3  fig.suptitle('Accuracy Score of Algorithms Predicting Category')
4  ax = fig.add_subplot(111)
5  plt.boxplot(results)
6  ax.set_xticklabels(names)
7  plt.show()
```

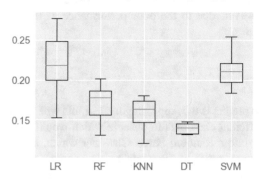

Fig. A.16 Confidence of Google App predictions

Developing a model based on categorical ratings of our full data yields two algorithms with the best performance: Logistic Regression and SVM with accuracy scores of 0.495. Furthering our analysis of the models trained on the full dataset, the testing accuracy for SVM edges out LR slightly with a score of 0.751–0.738. The results are shown in Fig. A.18. Although this is still too low for us to use reliably in the business setting, this model performs better than other models developed in this project.

Using a methodology and code for comparison developed by Dr. Richard Kunert[2], we were able to successfully compare use of a Linear Regression, Logistic Regression (one vs. rest and multinomial), and Ordered Logistic Regression from the "mord" package. The accuracy of the Ordered Logistic Regression exceeded the other models in its accuracy score, reaching 0.732. Again, the accuracy of this model is limited, and thus, no further exploration was performed.

Conclusion

It appears that none of the AI models yielded reliable results to predict our dependent variables. This may be the result of a number of factors, including the size of the training dataset, the quality of the data, and correlations of the data.

The lesson learned in this analysis is that there are no guaranteed results for predictive analyses. It is entirely possible to spend considerable resources (e.g., time

```
 1  #Create input features and target
 2  X = df_all_s[["Price_s","Installs_s","Reviews_s","Rating_s","Size_s","Android Ver"]]
 3  y = df_all_s.Category
 4  #Train-Test Split
 5  X_train, X_test, y_train, y_test = train_test_split(X, y, test_size=0.20, random_state=25)
 6
 7  from sklearn.ensemble import RandomForestClassifier
 8  #Train Model
 9  rfc = RandomForestClassifier(n_estimators=100).fit(X_train, y_train)
10  #Predict on Test Set
11  rfc_pred = rfc.predict(X_test)
12
13  #Find the accuracy score
14  print("Accuracy score:",accuracy_score(y_test, rfc_pred))
```

```
1  sns.barplot(x=feature_imp, y=feature_imp.index)
2  plt.xlabel("Feature Importance Score")
3  plt.ylabel("Features")
4  plt.title("Random Forest Importance Features")
5  plt.show()
```

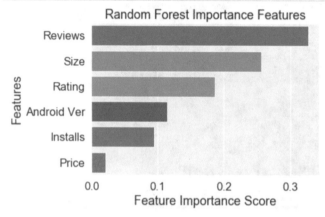

Fig. A.17 App feature importance using Random Forests

and money) conducting an analysis, only to obtain results with unreliable findings. This further emphasizes the importance of starting small, having a clear hypothesis, lots of data, and a good clean dataset.

References

1. 5 Steps of a Data Science Project Lifecycle.
2. Kaggle—Google Play Store Apps.
3. Modelling rating data correctly using ordered logistic regression.
4. Online Reviews Are Biased. Here's How to Fix Them.

```
 1  #Choose the Models to Try
 2  models = []
 3  models.append(('LR', LogisticRegression()))
 4  models.append(('RF', RandomForestClassifier()))
 5  models.append(('KNN', KNeighborsClassifier()))
 6  models.append(('DT', DecisionTreeClassifier()))
 7  models.append(('SVM', SVC()))
 8  # Set a seed
 9  seed = 123
```

```
 1  X = df_all_s[["Price_s","Installs_s","Reviews_s","Rating_s","Size_s","Android Ver"]]
 2  y = df_all.Category
 3  # Evaluate each model
 4  results = []
 5  names = []
 6  scoring = 'accuracy'
 7  for name, model in models:
 8      kfold = model_selection.KFold(n_splits=10, random_state=seed)
 9      cv_results = model_selection.cross_val_score(model, X, y, cv=kfold, scoring=scoring)
10      results.append(cv_results)
11      names.append(name)
12      msg = "%s: %f (%f)" % (name, cv_results.mean(), cv_results.std())
13      print(msg)
```

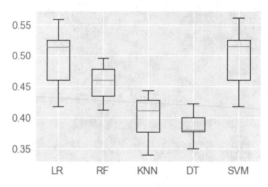

Fig. A.18 Confidence of Google App predictions

Migrating Python AI/ML Code to a Public Cloud

Setting Up AWS Cloud Machine and Running a Python File on It

Step 1: First start by setting up your Amazon Web Service (AWS) account. If you are doing it for the first time, as shown in Fig. A.19, then Amazon at the time of this writing was offering free compute tier for a year.

Step 2: Next, go to AWS console; this is a single pane of control that offers a glimpse into Elastic Compute Cloud (EC2), machine learning, containers, storage, database, and analytic services. Be mindful of checking your bill here often, as a running EC2 compute machine even if idle will incur charges which can add up to significant amounts of bills over time. Generally, Amazon's customer service is friendly and able to address unexpected charges as a one-time courtesy, but plan

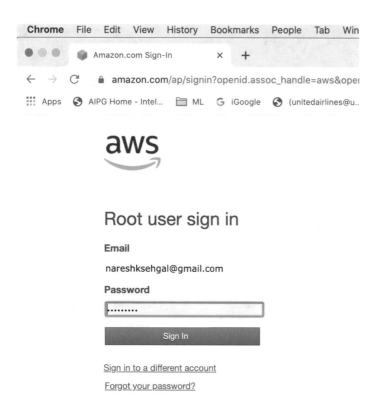

Fig. A.19 AWS sign-in window

ahead for what you need, use and its cost before launching new instances, storing data, or running tasks. Figure A.20 shows how the AWS console looks like.

Step 3: Click on EC2 and in that menu on key pairs. These are security keys to ensure communication between your desktop and server in the AWS cloud. You need to generate a new key pair, give it a name, and choose "pem" for use with OpenSSH. Clicking on Create Key Pair will result in successful creation of a new key pair. As you would notice in the next diagram, it is possible to maintain many separate keys for different types/purposes of cloud machines. You also have an option to import own keys under Actions menu, as shown in Fig. A.21. This is helpful when migrating AWS machines across accounts or users. Creating a new pair of keys will download a new .pem file to the local computer. Be careful to save it. Once lost, then access to any cloud assets created using it will be denied, and there is no alternative or replacement for the security keys.

Step 4: Now we are going to create and launch an AWS instance, in which your AI and ML tasks will be run using Amazon's cloud computing resources. Go to "EC2 Dashboard" on your AWS console and click on Launch Instance menu. It will bring up options. Each comes with a different cost structure for renting a part of Amazon's shared server in the cloud. If you are eligible for free tier, then use it, else

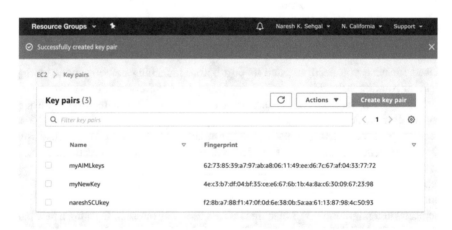

AWS Management Console

AWS services

Find Services
You can enter names, keywords or acronyms.

🔍 *Example: Relational Database Service, database, RDS*

▶ **Recently visited services**

▼ **All services**

　📋 **Compute**　　　　　　　　　　　　🧠 **Machine Learning**
　　　EC2　　　　　　　　　　　　　　　　　Amazon SageMaker
　　　Lightsail ⤴　　　　　　　　　　　　Amazon CodeGuru
　　　Lambda　　　　　　　　　　　　　　Amazon Comprehend
　　　Batch　　　　　　　　　　　　　　　Amazon Forecast
　　　Elastic Beanstalk　　　　　　　　Amazon Fraud Detector
　　　Serverless Application Repository　Amazon Kendra
　　　AWS Outposts　　　　　　　　　　Amazon Lex
　　　EC2 Image Builder　　　　　　　Amazon Machine Learning

Fig. A.20 AWS Cloud offerings

Resource Groups ∨ ★　　　　　　🔔 Naresh K. Sehgal ∨　N. California ∨　Support ∨

⊘ Successfully created key pair　　　　　　　　　　　　　　　　　　　　　✕

EC2 ＞ Key pairs

Key pairs (3)　　　　　　　　　　⟳　Actions ▼　**Create key pair**

🔍 *Filter key pairs*　　　　　　　　　　　　　　　　　＜ 1 ＞　⚙

☐	Name ▽	Fingerprint ▽
☐	myAIMLkeys	62:73:85:39:a7:97:ab:a8:06:11:49:ee:d6:7c:67:af:04:33:77:72
☐	myNewKey	4e:c3:b7:df:04:bf:35:ce:e6:67:6b:1b:4a:8a:c6:30:09:67:23:98
☐	nareshSCUkey	f2:8b:a7:88:f1:47:0f:0d:6e:38:0b:5a:aa:61:13:87:98:4c:50:93

Fig. A.21 AWS security keys

Amazon would have asked you for a credit card to pay for rental charges. Be mindful of what resources are being used. Be sure to stop the servers before you logout, else the cloud machines will keep running and incurring charges even after logging out of account. For our first assignment, we will pick the second machine as shown in Fig. A.22. It supports Python and is eligible for the AWS free tier.

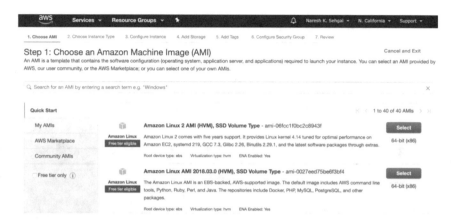

Fig. A.22 AWS sign-in window

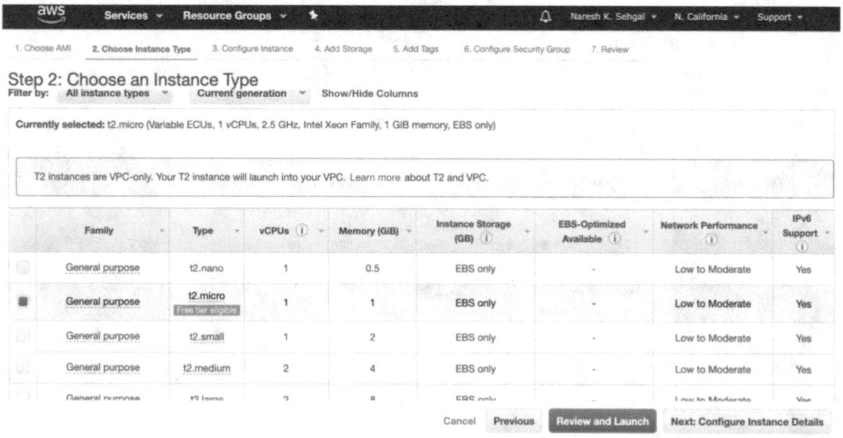

Fig. A.23 AWS Cloud machine choices

Step 5: In the next step, you will have to choose the type of instance. Since our data size and compute requirements are limited in the beginning, it is ok to go with t2.micro as shown in Fig. A.23. Simply select that and then click on "review and launch" button. At this point, your AWS charge meter has started unless you were eligible for the free tier.

Step 6: We will go with default choices for instance configuration, storage, and tags. Once you hit launch, AWS will ask you for the key pair to use for accessing this instance, as shown in Fig. A.24. It is important to save the "pem" file and even make a backup copy of it, as without it your instance and its content will be lost forever.

Step 7: Now you have a running instance. It may take a few seconds or minutes depending on the traffic at the AWS site. You can see running instances by clicking on the AWS console menu, as shown in Fig. A.25.

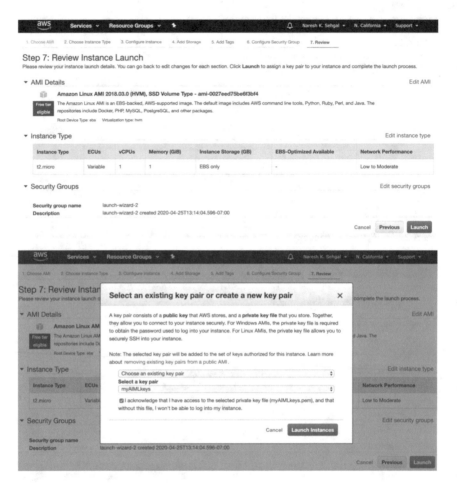

Fig. A.24 Launching an AWS Cloud machine

Step 8: Next we are going to "connect" to our AWS server, then transfer Python files and data to it as shown in Fig. A.26.

Step 9: Open a terminal window in which the following communication occurs to login using Linux SSH protocol and your security keys. The actual IP address for your AWS cloud machine will be different than what is shown below. It can be retrieved from the previous step. Next do a "sudo yum update," which will update all packages such as Python to the latest patches in AWS instance, as shown in Fig. A.27.

Step 10: Next step is to use the IP address of your EC2 instance, then copy your local data and any program files to the Amazon Machine Instance (AMI) in the cloud. Be sure to create a new folder AIMLbook on the AWS instance. You can open a new local terminal window and accomplish it with the following command,

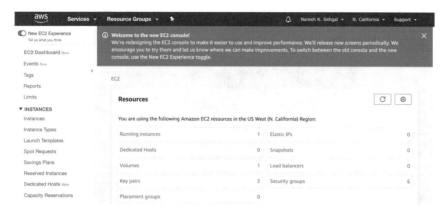

Fig. A.25 Launching an AWS Cloud machine

as shown in Fig. A.28, by giving your AMI's IP address after the ec2-user@IP:/home/ec2-user. You can get IP address from the AWS EC2 console.

- scp -i myAIMLkeys.pem *.* ec2-user@IP:/home/ec2-user/AIMLbook

Step 11: Now go back to the terminal window in which the connection to AMI cloud machine is open. Use Linux commands "ls" and "pwd" (for print working director) to see the current files, and their path to update it in your Python file. This will ensure that Google Play Store and user review csv files will be opened from the correct directory in the cloud, as shown in Fig. A.29.

Step 12: Next challenge is to run your Jupyter notebook interactive python files (say nb.ipynb) in a command line or script mode on the AWS. For this, you can convert to python text files (say nb.py), either through "save as" menu command or by using the command: Jupyter nbconvert to python nb.ipynb. More details are available here: https://github.com/Jupyter/nbconvert

Step 13: Before you run any Python file, make sure you have Python installed in your AWS cloud machine. Following series of commands will accomplish that:

- sudo yum install python36
- curl −0 https://bootstrap.pypa.io/get-pip.py
- python3 get-pip.py -user

Step 14: Next you need to ensure that all the right modules and libraries are available. It can be done by creating a file build.py with the following content in it starting with import command:

```
import pip
pip.main(['install','pandas'])
pip.main(['install','tweepy'])
pip.main(['install','seaborn'])
pip.main(['install','mord'])
pip.main(['install','scikit-learn'])
pip.main(['install','statsmodels'])
```

Fig. A.26 Connecting to an AWS Cloud machine

```
AIMLbook — ec2-user@ip-172-30-0-35:~ — ssh -i myAIMLkeys.pem ec2-user@ec2-3-101-41-139.us-west-1.compute.amazonaws.com —
nareshksehgal@machost AIMLbook % chmod 400 myAIMLkeys.pem
nareshksehgal@machost AIMLbook % ec2-3-101-41-139.us-west-1.compute.amazonaws.com
zsh: command not found: ec2-3-101-41-139.us-west-1.compute.amazonaws.com
nareshksehgal@machost AIMLbook % chmod 400 myAIMLkeys.pem
nareshksehgal@machost AIMLbook % ssh -i "myAIMLkeys.pem" ec2-user@ec2-3-101-41-139.us-west-1.compute.amazonaws.com
The authenticity of host 'ec2-3-101-41-139.us-west-1.compute.amazonaws.com (3.101.41.139)' can't be established.
ECDSA key fingerprint is SHA256:kLqhoQOs+aMmwVxxMVAIg85VlT00wCFi/shDtooqTZM.
Are you sure you want to continue connecting (yes/no/[fingerprint])? y
Please type 'yes', 'no' or the fingerprint: yes
Warning: Permanently added 'ec2-3-101-41-139.us-west-1.compute.amazonaws.com,3.101.41.139' (ECDSA) to the list of known hosts.

      __|  __|_  )
      _|  (     /   Amazon Linux AMI
     ___|\___|___|

https://aws.amazon.com/amazon-linux-ami/2018.03-release-notes/
4 package(s) needed for security, out of 7 available
Run "sudo yum update" to apply all updates.
[ec2-user@ip-172-30-0-35 ~]$
```

Fig. A.27 Remote terminal on an AWS Cloud machine

```
● ● ●                                                              Python — -zsh —
nareshksehgal@machost Python % scp -i myAIMLkeys.pem *.py ec2-user@3.101.41.139:/home/ec2-user/AIMLbook
PG.py
nareshksehgal@machost Python % scp -i myAIMLkeys.pem *.csv ec2-user@3.101.41.139:/home/ec2-user/AIMLbook
googleplaystore.csv
googleplaystore_user_reviews.csv
```

Fig. A.28 Copying data from a local terminal to an AWS Cloud machine

```
● ● ●        AIMLbook — ec2-user@ip-172-30-0-35:~/AIMLbook — ssh -i myAIMLkeys.pem ec2-user@ec2-3-101-41-139.us-west-1.compute.amazonaws.com
[ec2-user@ip-172-30-0-35 AIMLbook]$
[ec2-user@ip-172-30-0-35 AIMLbook]$ ls -lt
total 8872
-rw-r--r-- 1 ec2-user ec2-user 7669276 Apr 25 22:01 googleplaystore_user_reviews.csv
-rw-r--r-- 1 ec2-user ec2-user 1360155 Apr 25 22:01 googleplaystore.csv
-rw-r--r-- 1 ec2-user ec2-user   47912 Apr 25 22:01 PG.py
[ec2-user@ip-172-30-0-35 AIMLbook]$ pwd
/home/ec2-user/AIMLbook
```

Fig. A.29 Copying data from a local terminal to an AWS Cloud machine

Fig. A.30 Copying data from a local terminal to an AWS Cloud machine

Now execute it with "sudo python3 build.py" command, followed by "python3 PG.py" to run your Python file called PG.py. The results will show up on your AWS screen. If you wish to capture them in a text file, then use redirection with "python3 PG.py >& PG.txt" command.

It is interesting to see how the CPU and network traffic peaked in your AWS Cloud watch, as shown in Fig. A.30. Network peak came first as data was transferred from a local computer to the cloud. The CPU utilization peak came when the actual python file was run, as shown below.

Step 15: Be sure to transfer the resulting PG.txt file back to your desktop and stop your AWS Cloud machine, else bill will keep rising even if you disconnect the terminal connection.

Note 1: Every time you stop and start an Amazon instance, it will get a new and different IP address, so please note for connecting it to the remote terminal from your local machine.

Note 2: if you have a long-time running job on AWS and do not wish to stay connected with a live terminal then use a "Screen." It allows you to effectively add a new tab in the terminal window for running things in the background. Then you can disconnect, let cloud server run the job and reconnect to the screen for checking on it in between, or getting the final results later on. This can be done with the following five commands:

- sudo yum install screen, to install it for the first time only
- screen -S <my_screen_name>, to create a new screen
- launch your job &
- hit control+a+d to switch back to the main terminal
- screen -r <my_screen_name>, to return back to a screen
- screen -X -S <my_screen_name>, if you need to terminate it.

Appendix B: AI/ML for Letter Recognition

Using Python for Letter Recognition[2]

Background: Our objective is to identify each of a large number of black-and-white rectangular pixel displays as one of the 26 capital letters in the English alphabet.

Procedure:

1. The character images are based on 20 different fonts. Each letter within these 20 fonts was randomly distorted to produce a file of 20,000 unique stimuli.
2. Each stimulus is converted into 16 primitive numerical attributes (using statistical moments and edge counts).
3. These numerical attributes are then scaled to fit into a range of integer values from 0 through 15.

Step 1: Obtain Data

Letter Recognition Data Set from UCI https://archive.ics.uci.edu/ml/machine-learning-databases/letter-recognition/letter-recognition.data, as shown in Fig. B.1.

---

[2]*Acknowledgment*: This work is based on a class project done by Jammy Chan in a class taught by our lead author, Prof. Pramod Gupta, at UC Santa Cruz Extension in March 2020.

Fig. B.1 First half of
26 Roman alphabets, with a
method of geometrical
construction for large letters

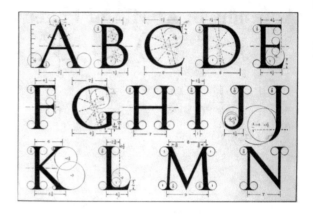

Step 2: Data Exploration and Attributes

We propose to start with a list of 16 attributes to categorize each of the given
alphabets. This is to calculate distinct features for each alphabet, which can be
used for training and recognition. The output of these will be a capital letter, e.g.,
26 values from A to Z.

1. x_box [horizontal position of box (integer)]
2. y_box [vertical position of box (integer)]
3. width [width of box (integer)]
4. height [height of box (integer)]
5. total_pixels [total # of pixels (integer)]
6. mean_x_pixels [mean x of on pixels in box (integer)]
7. mean_y_pixels [mean y of on pixels in box (integer)]
8. mean_x_variance [mean x variance (integer)]
9. mean_y_variance [mean y variance (integer)]
10. mean_xy_corr [mean x y correlation (integer)]
11. mean_x2y [mean of x * x * y (integer)]
12. mean_xy2 [mean of x * y * y (integer)]
13. x_edge [mean edge count left to right (integer)]
14. x_edgey [correlation of x-edge with y (integer)]
15. y_edge [mean edge count bottom to top (integer)]
16. y_edgex [correlation of y-edge with x (integer)]

Next, we introduce the concept of data balance or imbalance using Shannon's
entropy theorem. We calculate data imbalance for each of the 26 alphabets:

Balance = H/log k, such that 1 indicates perfect balance and 0 indicates imbalance,
where, k is the number of classes, C_i = size of cluster I, n = total size of samples, and
   H = −sum(C_i/n * log(C_i/n)) where i is from 1 to k.
As shown in the Fig. B.2 for letter "A," the balance value is rather high but not 1, due
   to differences between the left and right slope of the above figure.

Fig. B.2 Relative
measurements for letter "A"

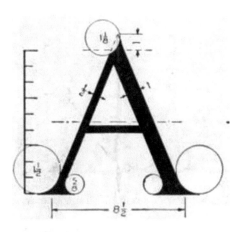

Fig. B.3 Data distribution and imbalance validation for all 26 letters

Figure B.3 shows the balance distribution for all 26 alphabets.

Step 3: Data Preprocessing

We perform splitting for different letter labels using their distinct features. These
feature values are normalized after standard scaling (Fig. B.4).

```
1  from sklearn.preprocessing import StandardScaler
2  # perform standard scaling
3  scaler = StandardScaler()
4  scaler.fit(letters_data)
5  letters_data_scaled = pd.DataFrame(scaler.transform(letters_data))
```

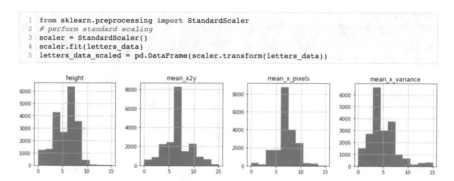

Fig. B.4 Distribution of 20,000 sample measurements across 16 numerical attributes

Step 4: Data Classification Models

We used the following list of AI classification models for our training set:

- Decision Tree
- Random Forest
- K-Nearest Neighbor
- Naïve Bayes
- SVC
- Logistic Regression

During the model evaluations, both hold out and K-fold validations were used. The drawback of hold-out validation is the accuracy which depends on the selection of the hold-out sets. For the given 20,000 observations, we used 80% for training and rest for testing. Random Forest was found as a suitable candidate based on the table as shown in Fig. B.5, with highest model accuracy, precision, and recall.

Step 5: Feature Importance and Confusion Matrix

Next we select Random Forest as our prediction model, and generated confusion matrix as shown in Fig. B.6. It shows an accuracy of 0.9652, precision of 0.9657, recall of 0.9651, while the F1 Score is 0.9652.

Then we rank the value of different features for recognizing the given letters. Results are shown in Fig. B.7. These imply that for letter recognition, instead of width and height of the letter, both x_edge and y_edge are the important features. In fact, the measurements of the letter edges (from left to right, top to bottom) provide the distinct characteristics of a letter.

```
1   # First constructing model class list
2   from sklearn.ensemble import RandomForestClassifier
3   from sklearn.tree import DecisionTreeClassifier
4   from sklearn.neighbors import KNeighborsClassifier
5   from sklearn.naive_bayes import GaussianNB
6   from sklearn.svm import SVC
7   from sklearn.linear_model import LogisticRegression
8   models = {}
9   model_eval = pd.DataFrame(columns = [])
10  seed = 123
11  numcpu = 4
12  models['RF'] = RandomForestClassifier(n_jobs=numcpu, random_state=seed, oob_score=True)
13  models['KNN'] = KNeighborsClassifier(n_jobs=numcpu, n_neighbors=5)
14  models['DT'] = DecisionTreeClassifier(criterion="gini", random_state=seed)
15  models['NB'] = GaussianNB()
16  models['SVM'] = SVC()
17  models['LOG'] = LogisticRegression(C=1, n_jobs=numcpu, random_state=seed, multi_class='ovr'
```

⇕	Model Name ⇕	Model Train Accuracy ⇕	Model Test Accuracy ⇕	Model Precision ⇕	Model Recall ⇕	Model F1 Score ⇕
0	RF	1.0000	0.9652	0.965747	0.965143	0.965221
4	SVM	0.9609	0.9445	0.946176	0.944307	0.944662
2	DT	0.9694	0.9432	0.943742	0.942981	0.943089
1	KNN	1.0000	0.8752	0.876008	0.874888	0.875070
5	LOG	0.7282	0.7210	0.724843	0.719716	0.720427
3	NB	0.6494	0.6485	0.662696	0.647375	0.644407

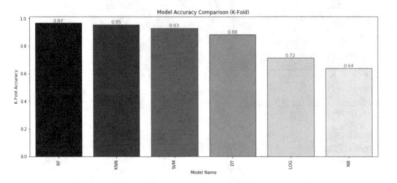

Fig. B.5 Comparison of different AI algorithms for 26 alphabets recognition task

Conclusion

Using classification model is one of the methods to perform letter recognition. More sophisticated deep learning methods like TensorFlow or neural networks should be used to classify/recognize complex objects. Deep learning may be required for tasks such as image recognition, audio real-time noise suppression.

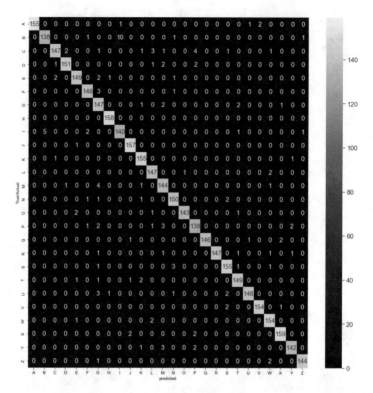

Fig. B.6 Confusion matrix for RF model

Fig. B.7 Importance of different geometric features for letter recognition

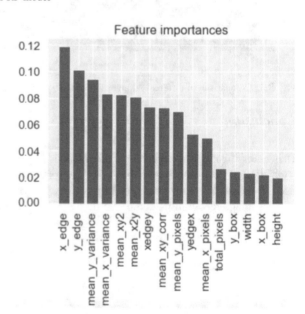

Future Work and Other Applications

A good extension for this work is to recognize cursive letters and even handwriting, by training with a sufficient dataset and then testing the programs. Appendix C has more applications along similar lines.

Migrating Code to Google Cloud Platform (GCP)

In this case, we will start by converting the Jupyter file to Python. Then upload this Python code to Github first, so it can be downloaded from there to a Cloud or share with other collaborators. If you have not used Github before, then go to https://github.com and set up an account, follow the instructions as shown in Fig. B.8, to upload the Python source files to Git.

Next step is to setup an account on Google Cloud and launch an instance. This is somewhat similar to what we did in the previous exercise on AWS. Details are shown in Fig. B.9.

Next step is to see your instance running, connect to it via a console using its external IP address, as shown in Fig. B.10.

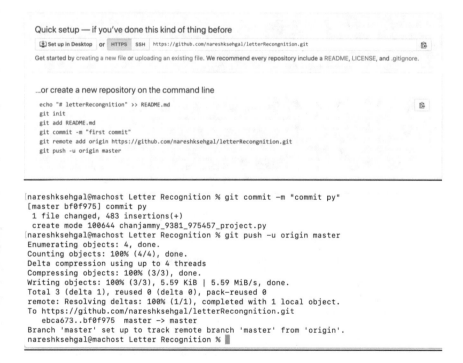

Fig. B.8 Uploading letter recognition code to Github

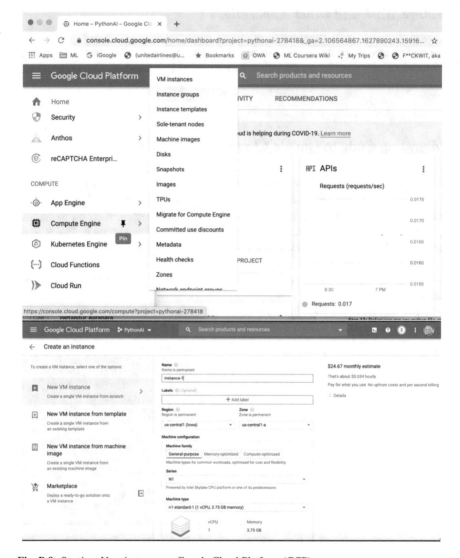

Fig. B.9 Starting. New instance on Google Cloud Platform (GCP)

Then we have three options to upload our Python files to the GCP server instance:

1. Via a download from the Github account, or
2. Directly from the local machine to the terminal window console, or
3. Via the storage bucket

Each of these has its own pros and cons. Using Github is good for programs that have many files, often in a directory structure. It also enables many different people to collaborate and contribute to the Git. Using upload from local machine to the

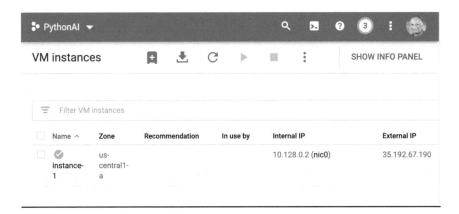

Fig. B.10 A running instance on GCP with its external IP for connecting to it

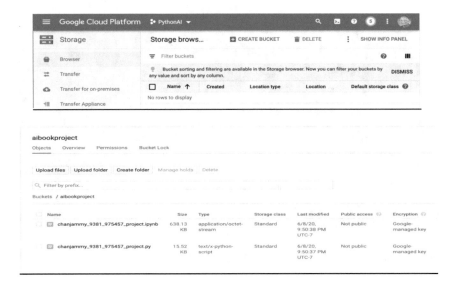

Fig. B.11 Creating a storage bucket and uploading files to it

terminal window works well for a single file, like what we have in this assignment. The storage bucket option is good for intermediate storage, especially if one will need the same file to run in different server instances, as each can connect directly to the bucket. We choose the storage bucket option for the purpose to continue this exercise, and Fig. B.11 shows this process.

Then we can transfer the files to console via the GCP bucket or Github or a directly upload to the GCP console. End results would look like as shown in Fig. B.12.

Fig. B.12 Seeing your Python files in the GCP console

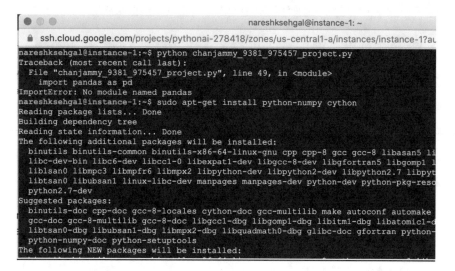

Fig. B.13 Install all libraries needed for running Python in a GCP server

As we can see in Fig. B.13, when we try to run "python" command with our .py file, it fails as module name pandas is not found. Hence, we install python-numpy using sudo (for super user access), and apt-get command, as shown below.

Now, we are ready to run our program from Appendix B1, or any other Python program in our GCP server instance.

Appendix C: Supervised Learning for MNIST Digits Dataset

Using Python ML Libraries for Letter Recognition[3]

Background: Our objective is to identify handwritten digits that are commonly used for training various image processing systems. Our source for these digits is MNIST (Modified National Institute of Standards and Technology). A sample is shown in

Fig. C.1 Sample images from MNSIT test dataset [1]

[3]*Acknowledgment*: This work is based on a class project done by Alex Wu in a class taught by our lead author, Prof. Pramod Gupta, at UC Santa Cruz Extension in August 2019.

Fig. C.1. The MNIST database contains 60,000 training images and 10,000 testing images. This database is also widely used for training and testing in the field of machine learning [1].

Procedure:

1. We start with exploratory data analysis (EDA) and preprocessing steps that include visualization and dimensionality reduction.
2. Next we will apply ML models from an existing Python library, i.e., Scikit-learn.
3. Then we will evaluate the results using a confusion matrix and hyperparameter tuning.
4. Lastly, we will explore more advanced techniques, e.g., neural networks.

Step 1: Download MNIST files with training set images and labels, as shown in Fig. C.2.

Step 2: Import all packages.

As a part of this exercise, we need to import several packages to evaluate different ML techniques on MNIST dataset. This is done as shown in Fig. C.3.

Step 3: Verify raw data.

Next we need to verify that the raw data is saved in a local folder, to avoid downloading the data every time. Then perform a limited exploratory data analysis on it, as shown in Fig. C.4 and visualize the digits as our inputs.

Step 4: Training with input dataset.

Next we create sub-datasets of large, medium, and small sizes. In most models, we will use the medium-sized datasets with 4000 training and 1000 test images, as shown in Fig. C.5.

For each sample, we see how many times a particular digit occurs. Ideally, this should be uniform, so all digits get an equal amount of training. We experiment with different samples and setting on the one where the heights of histograms are mostly uniform, as shown in Fig. C.6.

There are 4 files:

```
train-images-idx3-ubyte: training set images
train-labels-idx1-ubyte: training set labels
t10k-images-idx3-ubyte:  test set images
t10k-labels-idx1-ubyte:  test set labels
```

The training set contains 60000 examples, and the test set 10000 examples.

The first 5000 examples of the test set are taken from the original NIST training set. The last 5000 are taken from the original NIST test set. The first 5000 are cleaner and easier than the last 5000.

TRAINING SET LABEL FILE (train-labels-idx1-ubyte):

```
[offset] [type]          [value]          [description]
0000     32 bit integer  0x00000801(2049) magic number (MSB first)
0004     32 bit integer  60000            number of items
0008     unsigned byte   ??               label
0009     unsigned byte   ??               label
........
xxxx     unsigned byte   ??               label
```

The labels values are 0 to 9.

TRAINING SET IMAGE FILE (train-images-idx3-ubyte):

Fig. C.2 MNIST handwriting sample files [1]

```
 1  # from mnist_utils import load_data      # I have this file in same folder
 2  import matplotlib.pyplot as plt
 3  %matplotlib inline
 4  import seaborn as sns
 5  import random
 6  from scipy.stats import randint as sp_randint
 7  from time import time
 8  import datetime as dt
 9
10  import numpy as np
11  import pandas as pd
12  from sklearn.model_selection import train_test_split
13  from sklearn.decomposition import PCA
14  from sklearn.svm import SVC
15  from sklearn.naive_bayes import MultinomialNB
16  from sklearn.linear_model import LogisticRegression
17  from sklearn.linear_model import LogisticRegressionCV
18  from sklearn.ensemble import GradientBoostingClassifier
19  from sklearn import svm, metrics
20  from sklearn.model_selection import GridSearchCV, RandomizedSearchCV
21  from sklearn.metrics import roc_curve
22
23  from sklearn.neural_network import MLPClassifier
24  # from sklearn.metrics import confusion_matrix
25  from sklearn.metrics import accuracy_score, classification_report, confusion_matrix
```

Fig. C.3 Import all packages to be used in this notebook

```
 1  data_train= pd.read_csv("mnist_train.csv", header = None)
 2  data_test = pd.read_csv("mnist_test.csv", header = None)
```

```
 1  X_train = data_train.iloc[:,1:].values
 2  y_train = data_train.iloc[:,0].values
 3  print("Input train data:",X_train.shape); print("Output train data:", y_train.shape)
```

```
In [6]:     1  # meant to be some EDA
            2  len(X_train), type(X_train), X_train.shape, len(y_train)

Out[6]:  (60000, numpy.ndarray, (60000, 784), 60000)
```

```
 1  # plot 5 sampled images
 2  # permute_indices = np.random.permutation(train.shape[0])
 3  plt.figure(figsize=(17, 2.5))
 4  for i, (image, label) in enumerate(zip(X_train[0:5], y_train[0:5])):
 5      plt.subplot(1, 5, i + 1)
 6      plt.imshow(np.reshape(image, (28,28)), cmap=plt.cm.gray)
 7      plt.title('Training: %i\n' % label, fontsize = 20)
 8  plt.show()
```

Fig. C.4 Limited verification of input data correctness, before training begins

Step 5: Applying Naïve Bayes.

We start by applying a popular and efficient classifier, as shown in Fig. C.7. In machine learning, Naïve Bayes classifiers are a family of simple "probabilistic classifiers" based on applying Bayes' theorem with assumptions. With appropriate preprocessing, it is competitive in some domain, such as the text processing, with

```
 1  #%% # 2nd sample
 2  images_to_sample = 5000  #
 3  random_indices = random.sample(range(10000), images_to_sample)
 4
 5  X_train5K, X_test1K, y_train5K, y_test1K = \
 6  train_test_split(X_test10K[:images_to_sample,:], y_test10K[:images_to_sample,], test_size=0.20)
 7  print("--- Sample data shapes:", images_to_sample, X_train5K.shape, y_train5K.shape, \
 8        X_test1K.shape, y_test1K.shape, sep='\t')
 9  #%%
10  print('Train/test labels: %s : %s' % (np.unique(y_train5K), np.unique(y_test1K)))
11  print('Train/test class distribution:\n\t%s\n\t%s' % (np.bincount(y_train5K), np.bincount(y_test1K)))
12
13  if True:
14      plt.figure(figsize=(11.5,2.8))
15      plt.subplot(1,2,1)
16      plt.hist(y_train5K)
17      plt.ylabel("# Counts")
18      plt.xlabel("Digit")
19      plt.title("Sample #2: Train")
20      plt.subplot(1,2,2)
21      plt.hist(y_test1K)
22      plt.xlabel("Digit")
23      plt.title("Sample #2: Test")
24      plt.show()
```

Fig. C.5 Create smaller sample sets of MNSIT datasets

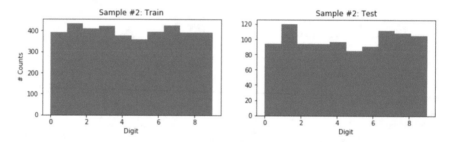

Fig. C.6 Representation of each digit in the sample data subsets

```
 1  # Naive Bayes
 2  #%% # xxxNB (GaussianNB), AdaBoost
 3  #from naive_bayes import MultinomialNB
 4
 5  for a in [0.001,0.01,0.1,1.0,5.0]:
 6      clf = MultinomialNB(alpha=a) # 0.001; def 1
 7      clf.fit(X_train5K, y_train5K)
 8      # MultinomialNB(alpha=1.0, class_prior=None, fit_prior=True)
 9      # print(clf.predict(X[2:3]))
10      #clf_acc_train =
11      #clf_acc_test =
12      print("  MultinomialNB alpha = {}, train/test acc: ".format(a))
13      print(clf.score(X_train5K[-1000:,], y_train5K[-1000:,]), \
14            clf.score(X_test1K, y_test1K))
```

Fig. C.7 Applying Naïve Bayes to MNIST dataset

more advanced methods including support vector machines [2]. It also finds application in automatic medical diagnosis [3].

Step 6: Logistic Regression (LR).

Next, we try LR, as shown in Fig. C.8. Although the initial accuracy was about 82%, a few try-and-error tunings made it reach 90%.

Step 7: Support vector machine (SVC).

Next we try SVC, also known as SVM in the Scikit-learn. Nonlinear kernels help to push the accuracy above 95%, as shown in Fig. C.9.

```
1  #%% ! SLOW ... ... ...
2  logisticRegr5K1 = LogisticRegression()     # take default solver (lin...')
3  logisticRegr5K1.fit(X_train5K, y_train5K) # using a larger dataset
4  #% % Predict for One Observation (image)
5  pred1_lr5K = logisticRegr5K1.predict(X_test1K[:10])
6
7  score1_train5K1K = logisticRegr5K1.score(X_train5K[-1000:,:], y_train5K[-1000:,])
8  score1_test5K1K = logisticRegr5K1.score(X_test1K[:1000,:], y_test1K[:1000,])
9  print(score1_train5K1K, score1_test5K1K)
10 # 1.0 0.824 % <-- not too bad
```

Fig. C.8 Applying Logistic Regression to MNSIT dataset.

```
1  # SVC from sklearn.svm import SVC
2  #%% SVC default, i.e. linear kernel
3  svc_def5K1K = SVC(kernel='linear')
4  svc_def5K1K.fit(X_train5K/256., y_train5K)        # make sure to norm
5  """
6  SVC(C=1.0, cache_size=200, class_weight=None, coef0=0.0,
7      decision_function_shape='ovr', degree=3, gamma='auto_deprecated',
8      kernel='linear', max_iter=-1, probability=False, random_state=None,
9      shrinking=True, tol=0.001, verbose=False)
10 """
11 #%% svc_pred = svc_def5K1K.predict(X_test)
12 svm_expected, start_time = y_test1K, time()
13 print('-- Starting SVC testing')
14 predicted = svc_def5K1K.predict(X_test1K/256.)
15 print('-- Completed SVC testing, duration {:.3f} seconds'.format(time()-start_time))
16 #%%
17
18 #sklearn.metrics
19 print("Classification report for SVC classifier %s:\n%s\n"
20       % (svc_def5K1K, classification_report(svm_expected, predicted)))
21
22 start_time = time()
23 trained = svc_def5K1K.predict(X_train5K[-1000:,:]/256.)
24 print('-- Completed SVC training evaluation, duration {:.3f} seconds'.format(time()-start_time))
25 # cm_tr = metrics.confusion_matrix(expected, predicted)
26 # print('Scikit-learn SVC ("linear kernel")\n   SVC Confusion matrix:\n%s' % cm)
27 print("   SVC training accuracy = {}".format(accuracy_score(y_train5K[-1000:,], trained)))
28 # train/test: 1.0 0.90 # not too bad
```

```
1  cm = confusion_matrix(svm_expected, predicted)
2  print('Scikit-learn SVC ("linear kernel")\n   SVC Confusion matrix:\n%s' % cm)
3  print("--> (Linear) SVC accuracy = {}".format(accuracy_score(svm_expected, predicted)))
4  print(classification_report(svm_expected, predicted))
```

```
Scikit-learn SVC ("linear kernel")
   SVC Confusion matrix:
[[ 86   0   2   0   0   3   1   0   2   0]
 [  0 118   1   0   0   0   0   0   0   1]
 [  4   2  80   2   3   0   0   0   3   0]
 [  0   2   0  86   0   2   0   2   2   0]
 [  1   0   0   1  88   0   1   0   0   6]
 [  3   1   0   3   4  71   1   0   2   0]
 [  1   0   3   0   1   2  84   0   0   0]
 [  0   0   1   0   2   0   0 105   0   4]
 [  1   3   1   2   1   5   1   0  94   0]
 [  1   0   1   0   4   0   0   2   2  95]]
--> (Linear) SVC accuracy = 0.907
```

Fig. C.9 Applying SVM to MNIST dataset

```
 1  ############## Classifier with better params ##############
 2  #from sklearn.svm import SVC
 3  # Create a classifier: a support vector classifier
 4  param_C = 5
 5  param_gamma = 0.05
 6  clf5_SVM5K = SVC(C=param_C, gamma=param_gamma)
 7  print(clf5_SVM5K) # model info
 8  if True:
 9      #We learn the digits on train part
10      start_time = time() # time.time()
11      print('-- Starting SVM learning at system time {:.6f}'.format(start_time))
12      # clf5_SVM5K.fit(X_train08K, y_train08K)
13      clf5_SVM5K.fit(X_train5K/256., y_train5K)
14      end_time = time()
15      print('-- Completed SVM learning with duration {:.3f} seconds'.format(time()-start_time))
16      # elapsed_time= end_time - start_time
17      # print('Elapsed learning {}'.format(str(elapsed_time)))
18  #
19  #% %
20  """ X_train5K
21  -- Starting SVM learning at system time 1567894504.397274
22  -- Completed learning with duration 44.681 seconds
23  """
24  ############################################################
```

Fig. C.10 Applying SVM to MNIST datasets

```
 1  # random forest
 2  ### RF via sklearn: https://docs.w3cub.com/scikit_learn/modules/generated/sklearn.ensemble.randomforestclassifier/#
 3  from sklearn.ensemble import RandomForestClassifier
 4
 5  clf_rf5K1K = RandomForestClassifier(n_estimators=100, max_depth=5,
 6                                       random_state=5000)
 7  clf_rf5K1K.fit(X_train5K, y_train5K)
 8  print(clf_rf5K1K.feature_importances_)
 9  ###
10  trainingOutputs = clf_rf5K1K.predict(X_train5K[-1000:,:]) # trainingSetSize
11  testOutputs = clf_rf5K1K.predict(X_test1K)
12
13  # not yet evaluation
14  ### gridSearch: https://e-string.com/articles/random-forests-scale/
15  model = RandomForestClassifier()
16  parameters = [{"n_estimators": [300, 550, 800], \
17                  'max_depth': [30,40,50], \
18                  'random_state': [5000]}]
19  clf_gridRF5K1K = GridSearchCV(model, parameters, verbose=5, n_jobs=8)
20  clf_gridRF5K1K.fit(X_train5K, y_train5K)
21  clf_gridRF5K1K.score(X_test1K, y_test1K)
22
23  print("Best Params: " + str(clf_gridRF5K1K.best_params_))
24  print("Best Score: " + str(clf_gridRF5K1K.best_score_))
25  print("Best Estimator: " + str(clf_gridRF5K1K.best_estimator_))
```

Fig. C.11 Applying Random Forest to MNSIT datasets

Step 8: Applying nonlinear (RBF) kernel (Fig. C.10).

Step 9: Random Forest (RF).

Lastly, we try RF as shown in Fig. C.11, which yielded a match with 92.6% accuracy that met our threshold of expectations. While, gradient boosting only gave correct results 82.7% of the times, which is below our expectations.

Conclusions: Out of all ML methods, SVM is the best, as shown by the following heatmap matrix in Fig. C.12.

Next we start with a multi-layer perceptron (MLP) model, as shown in Fig. C.13.

An artificial neural network [4] is an interconnected group of nodes, inspired by neurons in a brain. Each circular node represents an artificial neuron, and an arrow represents a connection from the output of one artificial neuron to the input of another. Our implementation of MLP is shown in Fig. C.14. Our MLP trains on two arrays, array X of size (samples, features), which holds the 5000 training

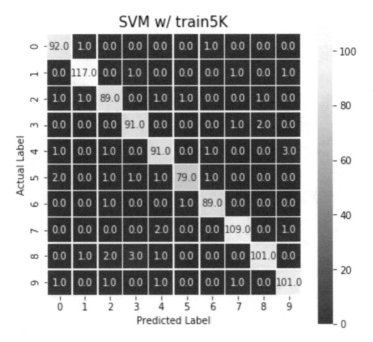

Fig. C.12 Most matches were found using SVM

Fig. C.13 Representation of an artificial neural network

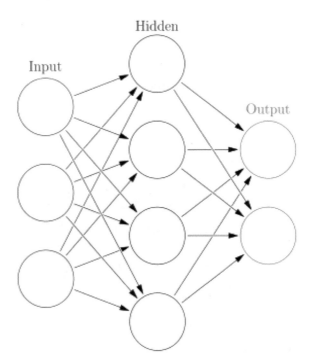

```
1   # NN
2   # https://scikit-learn.org/stable/auto_examples/neural_networks/plot_mnist_filters.html#sphx-glr-auto-examples-neur
3   #from sklearn.neural_network import MLPClassifier
4   # mlp = MLPClassifier(hidden_layer_sizes=(100, 100), max_iter=400, alpha=1e-4,
5   #                     solver='sgd', verbose=10, tol=1e-4, random_state=1)
6   mlp_5K1K = MLPClassifier(hidden_layer_sizes=(50,), max_iter=1600, \
7                            alpha=1e-4, solver='adam', verbose=6, tol=5.0e-5, \
8                            learning_rate='invscaling', \
9                            learning_rate_init=.05, random_state=5000)
10  # --> MLP test acc.: 0.89
11  # https://scikit-learn.org/stable/auto_examples/neural_networks/plot_mlp_training_curves.html#sphx-glr-auto-example
12  """                      # ='invscaling' or adaptive
13  MMLPClassifier(activation='relu', alpha=1e-05, batch_size='auto', beta_1=0.9,
14                 beta_2=0.999, early_stopping=False, epsilon=1e-08,
15                 hidden_layer_sizes=(50,), learning_rate='constant',
16                 learning_rate_init=0.05, max_iter=1000, momentum=0.9,
17                 n_iter_no_change=10, nesterovs_momentum=True, power_t=0.5,
18                 random_state=5000, shuffle=True, solver='sgd', tol=5e-05,
19                 validation_fraction=0.1, verbose=6, warm_start=False)
20  ...
21  Training loss did not improve more than tol=0.000050 for 10 consecutive epochs. Stopping.
22      Simple (1-layer) MLP train acc.: 1.00
23  --> MLP test acc.: 0.91 <-- same acc as last set of param.
24  """
25  #% %
26  if True:
27      mlp_5K1K.fit(X_train5K/256., y_train5K)  # <-- training
28      mlp_score_train = mlp_5K1K.score(X_train5K[-1000:,:]/256.0, y_train5K[-1000:,])
29      # print("Training set score: %.2f" % mlp_5K1K.score(X_train5K[-1000:,:]/256., y_train5K[-1000,]))
30      mlp_score_test = mlp_5K1K.score(X_test1K, y_test1K)
31      print("    Simple (1-layer) MLP train acc.: %.2f" % mlp_score_train)
32      print("--> MLP test acc.: %.2f" % mlp_score_test)
33
34  # 0.98 0.91
```

Fig. C.14 MLP-based training for MNIST classification

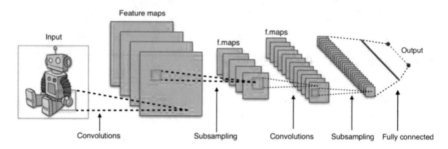

Fig. C.15 A typical CNN architecture [5]

```
In [ ]: # hide the 2nd convolution layer
        class MNIST_CNN1(nn.Module):
            def __init__(self):
                super(MNIST_CNN1, self).__init__()
                self.conv1 = nn.Sequential(       # input shape (1, 28, 28)
                    nn.Conv2d(
                        in_channels=1,            # input height
                        out_channels=32,          # n_filters
                        kernel_size=5,            # filter size
                        stride=1,                 # filter movement/step
                        padding=2,                # if want same width and length of this image after Conv2d, padding=(kerne.
                    ),                            # output shape (16, 28, 28)
                    nn.ReLU(),                    # activation
                    nn.MaxPool2d(kernel_size=2),  # choose max value in 2x2 area, output shape (16, 14, 14)
                )
                ...
                self.conv2 = nn.Sequential(       # input shape (16, 14, 14)
                    nn.Conv2d(32 , 32, 5, 1, 2),  # output shape (32, 14, 14)
                    nn.ReLU(),                    # activation
                    nn.MaxPool2d(2),              # output shape (32, 7, 7)
                )
                ...
                self.out = nn.Linear(32*14*14, 10)
                # self.out = nn.Linear(32 * 7 * 7, 10)   # fully connected layer, output 10 classes

            def forward(self, x):
                x = self.conv1(x)
                # x = self.conv2(x)
                x = x.view(x.size(0), -1)         # flatten the output of conv2 to (batch_size, 32 * 7 * 7)
                output = self.out(x)
                return output, x    # return x for visualization
```

Fig. C.16 CNN application for MNIST

samples represented as floating point feature vectors, and array Y of size (samples), which holds the target values (class labels) for the training samples. After ten iterations, the program stops when training error did not reduce any further significantly. This result is better than all the previous methods we had considered so far.

However, there is still a room for improvement, so next we try convolution neural network (CNN) [5]. The name "convolutional neural network" indicates that the network employs a mathematical operation called convolution [6]. Convolution is a specialized kind of linear operation. Convolutional networks are simply neural networks that use convolution in place of general matrix multiplication in at least one of their layers (Fig. C.15).

CNNs are often used in image recognition. Our code for CNN is shown in Fig. C.16.

This helps to further improve our results to 0.992, which is good. However, based on all the methods we tried so far, MLP is the best suited for this problem.

References

1. https://en.wikipedia.org/wiki/MNIST_database.
2. http://yann.lecun.com/exdb/mnist/index.html.
3. https://scikit-learn.org/stable/supervised_learning.html.
4. https://en.wikipedia.org/wiki/Artificial_neural_network.
5. https://en.wikipedia.org/wiki/Convolutional_neural_network.
6. https://en.wikipedia.org/wiki/Convolution.

Appendix D: Points to Ponder

Chapter 1: Points to Ponder

1. What is the difference between AI and ML?

   - Artificial intelligence (AI) is the broader concept of machines acting in a smart manner, such as for playing chess. Machine learning (ML) is subdomain of AI based on an idea that given sufficient amount of data, machines can learn the rules without explicit instructions or programming. Thus, ML is one of the ways for machines to develop AI capabilities.

2. What types of problems are suitable for unsupervised learning to solve?

   - In unsupervised learning, machines find meaningful relationships and patterns in a given dataset. It is useful when labeled data is not available, in cases such as finding groups or clusters, extracting generative features, or for exploratory purposes.

3. What are the advantages and concerns of doing ML in a public cloud?

   - Cloud-based machine learning is good for applications that need to analyze large quantities of data. If that data is coming from many different sources over a period of time, then Cloud storage is an attractive place to store that data for long term. An example of such cases is Internet of Things (IoT) and healthcare diagnostics. However, some businesses have privacy and security concerns about storing their data in a public cloud, for example, all medical data storage and transmission must abide by HIPPA protocols.

Chapter 2: Points to Ponder

1. How can one measure the correctness of an ML algorithm?

   • Once an ML model has been trained, one simple way to measure its accuracy is to use the model to predict answers with test dataset and compare the results to the actual answers. It is important to use an independent test set, and not the training dataset. Various ML techniques can be compared using statistical tests, for example, root mean square error (RMSE). Then a confusion matrix (aka error matrix) is constructed. The name comes from the fact that it is important to check if an ML system is confusing between classes, i.e., mislabeling one as another. It is a tabular layout presenting visualization of the performance of an algorithm. For unsupervised learning, it is called a matching matrix. Each row of the matrix represents the instances in a predicted class, while each column represents the instances in an actual class.

2. How can one measure the performance of an ML algorithm?

   • Performance is a time-based measure and different from accuracy, which measures correctness. Ideally, one desires both accuracy and faster runtime performance in any algorithm. Since different algorithms perform differently for a given dataset, it is important to measure runtime for performance comparisons. Furthermore, performance of an algorithm depends on whether it was run on a CPU, GPU, TPU, or any other specialized hardware. Choice of operating system (e.g., windows vs. Linux), software libraries, and profiler also impact the runtimes. It is important for making sure that the entire runtime environment is considered for a fair apples-to-apples comparison in selecting an algorithm.

3. What is the meaning of over-fitting and why it is not desirable?

   • Over-fitting is a modeling error, which occurs if a function is defined to closely fit to a limited set of data points during the training phase. It makes a model conform to mimic a dataset that may not be fully representative of other data points that the model may encounter in future. Thus, it may result in substantial errors when the model is used for inference.

Chapter 3: Points to Ponder

1. What are the differences between machine learning (ML) and deep learning (DL)?

   • Recall that ML is a branch of AI that can self-learn based on a given dataset and improve its decisions over time without human intervention. Similarly, DL is a sub-branch of ML, which can be applied to extremely large datasets. The word deep comes from multiple layers in an artificial neural network (ANN).

2. What are the advantages of DL?

- DL breaks down a complex problem in stages and uses layered solution processes to create an artificial human brain-like structure to make intelligent decisions. Each layer in DL system represents a stage, where parameters can be tuned for making complex decisions, e.g., for Netflix to decide which movie to suggest next based on past viewing habits of a subscriber.

3. What are the drawbacks of DL?

- Similar to everything else in life, DL also has its advantages and disadvantages. DL requires a very large amount of data to perform better than other AI solutions. A DL model is computationally expensive to train due to complex data model with many variables. It may require expensive GPUs and other specialized hardware machines. A DL system is hard to debug, in case of any errors in the output.

Chapter 4: Points to Ponder

1. If a cloud service provider wants to offer ML at IaaS layer, what will be the features of such a service?

- A user of IaaS wants to avoid capital expenditure and uses cloud facility to pay for it on per use basis. IaaS user is mostly concerned with the quality of service in terms of a cloud server's compute, memory, and network latencies. ML in IaaS can be used to track the response time for users' hosted applications and enabling them to schedule tasks so as to maximize their compute efficiency. Any idle servers can be shut down and workloads consolidated to maximize the utilization of the running servers. New servers can be started up as users' workload demands increases. ML can be helpful for tracking usage metrics, predicting costs when scalability is needed to maintain a constant QoS, etc.

2. If a cloud service provider wants to offer ML at PaaS layer, what will be the features of such a service?

- Users of PaaS is mostly concerned about the specific tasks related to their hosted services, such as database I/O transactions and customer activities on their hosted websites. Examples of PaaS are a Python-integrated development environment or MatLab tools and facilities. ML can be helpful for generating metrics related to end-user experiences such as search times for catalog items, wait times, and optimizations related to other services such as payments.

3. If a cloud service provider wants to offer ML at SaaS layer, what will be the features of such a service?

- A user of SaaS, such as Netflix or Salesforce, may be mostly concerned about the statistics and preferences related to their applications' end-users. These services need to be provided on an expeditious basis. ML can be helpful to track the SaaS customers' preferences and make suggestions based on AI models and past behavior to suggest new movies, etc.

Chapter 5: Points to Ponder

1. What is the need for data clean-up before feeding it to ML algorithms?

   - Goal of data clean-up is to identify and remove any duplicate or erroneous data elements, to create a reliable dataset. This improves the quality of input data for the ML model training. It results in accurate decision-making during ML inference phase.

2. Why is feature scaling necessary?

   - ML algorithms are sensitive to the relative scale of features, which are the input values used for training a model. Feature scaling is needed for ML algorithms that calculate and use distances between data. Examples of such algorithms include KNN, K-Means, PCA, etc. If input data is not scaled or normalized, a feature with higher values will dominate ML parameter settings and thus skew the model results.

3. What is the impact of multicollinearity?

   - Multicollinearity refers to a phenomenon in which one variable in a *dataset* can be linearly predicted from the others with a substantial accuracy. This can cause ML model's coefficients to change erratically in response to small changes in the input data. It may also lead to over-fitting.

Chapter 6: Points to Ponder

1. Why is it important to separate out private data for encryption?

   - User data may be of two types: usually or freely available, vs. sensitive or private data. The first category consists of things like name, photograph, location, etc. This information may be easily available to anyone via an Internet search. The second category may have some personal information such as date of birth, bank account number, and salary. Loss of this information may cause irreparable harm. Someone may impersonate to withdraw of money from the bank. Thus, it is important to keep it secret by encrypting it. Another reason to separate these two categories is the cost of encryption. It takes time and computational power to do encryption and decryption, so such care should be applied to sensitive information only.

2. How does a public cloud make security issues more complicated than an enterprise or private cloud?

- Information storage in a public cloud is not directly under a user's control. The system administrator may make additional backup copies too. Therefore, the possibility to hack private information during transmission or storage is higher, or perceived to be higher, in a public cloud.

3. What is the need to encrypt data during transmission from a user site to the public cloud data center?

- Data packets may travel through one or more public nodes during transmission from the user site to a public cloud; hence, it is better to encrypt the sensitive data before sending it and maintain the keys in a different location.

4. Are there any runtime concerns for data and programs in a public cloud?

- We talked about encrypting data at rest and during transition to a public cloud. In addition, the servers are shared among many users in a multi-tenant mode of operation. This is akin to a hotel renting rooms to different guests on the same floor. If the walls between the rooms are thin, then other guests can hear the conversation of their neighbors. Thus, it is better to consider in-memory encryption of sensitive data as well as to avoid a noisy neighbor problem.

5. Traditional AI has two distinct phases for training and inference. How does this need to change for security applications?

- In traditional AI, training data is labeled and then classified for the learning system. This works well for static situations, such as differentiating between the pictures of dogs and cats. However, most security situations are unique as new virus patterns emerge or hackers find new vulnerabilities in existing code bases. Thus, the training and inference phases need to be interleaved. Human element is also needed as a previously safe situation may be marked unsecured due to new findings. Hence, AI for security is still an evolving research area.

Chapter 7: Points to Ponder

1. How has the spread of Internet enabled more data generation?

- Internet connects computers and devices across geographical distances, yet enables them to communicate data and decisions in real-time. This has led to information accumulation at a rate unprecedented in the human history. In a single day, more data is generated than in hundreds of years preceding the advent of Internet.

2. Why is it important to consider distributed decision-making in cloud?

- While data generation is a global phenomenon, decision making is still centralized mostly due to federated models of learning and authority concentration. However, the need to make local decisions is imperative in the interest of timeliness. An example is a self-driving car, which cannot wait for a cloud server to decide whether the obstruction in front is a real person or shadow of a tree on the camera. Time constant required here is in the order of milliseconds. In such cases, a human or local computer must make a split-second decision, then the results can be sent back to the central computing for future decision improvements.

3. Is data-based training a one-time activity or iterative?

   - In many situations, such as cancer detection in early stage patients, it is hit-n-miss game for the doctors. If this task is relegated to an AI/ML system, then initial training set is not sufficient and the parameters for decision-making need to be updated regularly and iteratively using a feedback system.

4. Would Hadoop be still popular if it was not open-sourced?

   - In the opinion of authors, answer is no as its adoption beyond initial usage by Yahoo would have been limited. An example if Google's page rank algorithm, which makes its search engine results more useful than competitors. By contributing Hadoop to the open-source community, Yahoo may have limited their financial profits, but have contributed to the growth of cloud computing and large dataset processing.

5. How can the inference results of NLP feedback into the training phase?

   - A part of the initial training data is not exposed to the NLP system. Based on the initial training, the system is presented with new input data, and its results are compared with the known good output. Any delta or error between the expected and predicted results are presented as a feedback to the NLP system, for improving its parameters. This process is repeated until the error rate reduces to an acceptable level.

Chapter 8: Points to Ponder

1. Discuss if we are at the cusp of a medical revolution with confluence of AI and cloud?

   - Healthcare is at an inflection point due to a growing population, shortage of medical practitioners and limited public resources, such as hospitals. With lots of new devices and data availability, medical science is evermore ready for individualized treatments instead of one size fits all. Telemedicine is already being practiced during Covid times to minimize risk and improve efficiency. However, most of the patients' data is still locked up in different hospitals' databases due to HIPPA regulations and perceived risk of moving it to a public cloud. Next obvious steps are to bring this data in a secure manner to the clouds and bring together the power of AI to assist physicians to identify patients' condition and recommend appropriate treatments.

2. What are the advantages of TCD over MRI and CT scans for brain disease diagnosis?

- A transcranial Doppler device uses 2–2.5 MHz ultrasound waves, which are similar to the devices used for scan for an unborn baby in her mother's womb. Thus, energy exposure at that frequency is safe for adults too. CT scans use X-rays which can deliver harmful radiation if a patient is exposed often or for over longer durations of time. MRI tends to be expensive and takes long time to conduct, as compared to a quick TCD exam.

3. Why some companies prefer a multi-cloud solution?

- Many cloud customers want to avoid a vendor lock-in, for the fear of future price raises and limited capabilities as compared to other cloud vendors. A multi-cloud solution avoids the above issues, as well as enables a business to support its customers who may prefer one cloud vendor or another. An example is a healthcare provider in the cloud, whose customers are hospitals with their on existing agreements or preference for difference Cloud Service Providers (CSPs). In this case, healthcare provider will have a hard time convincing a hospital to move its data from one cloud to another, and must adopt a multi-cloud solution. Such a solution may need to cross organizational boundaries, and use the Internet to access storage in one cloud, while algorithms may be running in another cloud.

Chapter 9: Points to Ponder

1. Can faster AI and ML solutions be used to make up for paucity of data?

- In some cases, synthetic data can be generated from some given seeds. An example is the training of an automated driving vehicle, where additional traffic conditions can be simulated. However, in some other cases, such as effects of a cancer control medicine, there is no good substitute for actual human or animal trials. Synthetic data does not apply to modeling and predicting stock market behavior too, with many different parameters and participants.

2. Why AI training computation needs are higher than for inference?

- AI training generally deals with larger datasets, and neural networks need time to determine parameters. Sometimes backpropagation is needed to correct any errors between computed and expected results. Large datasets can be handled by many parallel machines, as described in Chap. 9.

3. How can an FPGA-based AI and ML solution outperform GPU?

- It depends on whether the model requires more low-level compute or software-based logic. In the latter case, since FPGA can minimize the effects of OS and driver layers by mapping AI algorithms directly to hardware. In that case it will be faster than a GPU.

4. Why does an ASIC solution offer higher performance than FPGA?

- If hardware mapping of an FPGA results in an inefficient mapping of the program logic to gates placed or routed far from each other, then a significant amount of time will be spent in electrical signal transfers. An ASIC on the other hand uses custom place-and-route solutions to build a special purpose hardware, which may outperform GPUs and FPGAs. Only downside is the design cost that needs to be amortized over a large number of units.

5. How a high bandwidth memory may reduce the need for communication between chips?

- A high bandwidth memory (HBM) is akin to having a large cache in a CPU, where chunk of data is fetched before it is needed. Thus, more time can be spent to compute and less on the communication between the chips. Often HBMs can be shared across many compute units, as shown with hardware AI accelerators in Chap. 9.

Appendix E: Questions for Practice

1. What is the life cycle of a data science project?
2. Compare R and Python.
3. What is machine learning? Where and why to use machine learning?
4. What is the difference between data mining and machine learning?
5. What are the four analytics in data science?
6. What are feature vectors?
7. What is the difference between normalization and standardization?
8. What is normalization/scaling in machine learning and why do you perform?
9. What is the difference between regularization and normalization?
10. What is supervised and unsupervised learning? Give concrete examples.
11. How do you handle missing data? What imputation techniques do you recommend?
12. What is confusion matrix and why do you need it?
13. What do you mean by Type I and Type II errors?
14. What is ROC curve and what does it represent?
15. Explain what precision and recall are. How do they relate to the ROC curve?
16. What is imbalance of data? How do we fix it?
17. You are given a dataset on cancer detection. You have built a model and achieved an accuracy of 97%. Why should not be happy? What can you do about it?
18. Explain the handling of missing or corrupted values in the given dataset.
19. Explain what a false positive and a false negative are. Provide examples when false positives are more important than false negatives, false negatives are more important than false positives.
20. What is the difference between one hot encoding and label encoding? Which one result in dimensionality increase?
21. What is binarizing of data? How to binarize?
22. What is the difference between classification and regression?
23. What is clustering and how is it different from classification?
24. How do you select important features in a dataset? Explain your methods.

© The Author(s), under exclusive license to Springer Nature Switzerland AG 2021
P. Gupta, N. K. Sehgal, *Introduction to Machine Learning in the Cloud with Python*,
https://doi.org/10.1007/978-3-030-71270-9

25. What is dimensionality reduction? Why it is performed and what techniques do you know?
26. What is principal component analysis (PCA)? Explain where you would use PCA.
27. Do you know/used data reduction techniques other than PCA? What do you think of step-wise regression? What kind of step-wise techniques are you familiar with?
28. What are the assumptions required for linear regression? What if some of these assumptions are violated.
29. What is multicollinearity? How to remove multicollinearity?
30. Is it possible to capture the correlation between continuous and categorical variable. If yes, how?
31. What is curse of dimensionality?
32. How to check if the regression model fits the data well?
33. Linear regression model is generally evaluated using adjusted R^2. How would you evaluate a logistic regression model?
34. How do you know if one algorithm is better than other?
35. What assumptions are made about the attributes in Naïve Bayes method and why?
36. What is Laplacian correction and why it is necessary?
37. What is over-fitting in ML? What are the different methods to handle it?
38. Is a high variance in a data good or bad?
39. If your data is suffering from high variance, how would you handle it?
40. What is ensemble learning? Why and when it is used?
41. What is a decision tree? Describe a strategy that can be used to avoid overfitting in a decision tree.
42. How can a decision tree be converted into a rule set? What are the advantages of the rule set representation over the decision tree representation?
43. When do you use random forests vs. SVM and why?
44. What are Kernels in SVM? List popular kernels used in SVM.
45. What is imbalance data and what are the various ways to handle it?
46. Take the Boston housing data that is available in sklearn. We want to predict median house price. Select the features from the given data. Arrange the features in the order of their importance. Which variables have a strong relation to median price? Divide the data into training and testing. Fit a linear model. Plot the predicted values of the test data.
47. What does K stand for in K-nearest neighbor algorithm? What are the different methods to determine the optimal value of K in K-means algorithm?
48. What is gradient descent method? Will gradient descent methods always converge to the same solution?
49. What impurity measures do you know?

Glossary

Accuracy Accuracy is a metric to indicate goodness of ML classification model.

Activation function *Activation function* is a nonlinear function applied to the weighted sum of the inputs of a *neuron* in a *neural network*. The presence of activation functions makes neural networks capable of approximating virtually any function.

Artificial intelligence (AI) AI is that part of computer science which aims to build machines capable of doing human-like tasks such as decision-making, object classification and detection, speech recognition, and translation.

Autoencoder An *autoencoder* is a *neural network* whose goal is to predict the input itself, typically through a *bottleneck* layer somewhere in the network. By introducing a bottleneck, the network is forced to learn a lower-dimensional representation of the input, effectively compressing the input into a good representation. Autoencoders are related to *principal component analysis* and other *dimensionality reduction* and *representation learning* techniques but can learn more complex mappings due to their nonlinear nature.

Bag of words *Bag of words* is a method of feature engineering for text documents. According to the bag of words approach, in the feature vector, each dimension represents the presence or absence of a specific token in the text document. Therefore, such an approach to representing a document as a feature vector ignores the order of words in the document. However, in practice, bag of words often works well in document classification (see Token later).

Bagging Bagging or bootstrap averaging is a technique where multiple models are created on the subset of data, and the final predictions are determined by combining the predictions of all the models. It is designed to improve the stability and accuracy of ML algorithms. It also reduces variance and helps to avoid over-fitting (see over-fitting later).

Baseline A *baseline* is an algorithm, or a heuristic, which can be used as a *model*. It is often obtained without using machine learning by using the most simplistic engineering method.

© The Author(s), under exclusive license to Springer Nature Switzerland AG 2021
P. Gupta, N. K. Sehgal, *Introduction to Machine Learning in the Cloud with Python*,
https://doi.org/10.1007/978-3-030-71270-9

Batch *Batch*, or mini-batch, is the set of examples used in one iteration of model training using *gradient descent.*

Bias Bias is an error from erroneous assumptions in the learning algorithm. High bias can cause an algorithm to miss the relevant relations between features and target outputs (under fitting).

Bias–variance tradeoff The *bias–variance tradeoff* is the property of a set of *predictive models*, whereby models with a lower *bias* in parameter estimation have a higher *variance* of the parameter estimates across samples, and vice versa. The bias–variance problem is the conflict in trying to simultaneously minimize these two sources of error that prevent *supervised learning* algorithms from generalizing beyond their *training set.*

Big data Big data is a term that describes the large amount of data—both structured and unstructured. But it is not the amount of data that is important. It is how organization uses this large amount of data to generate insights.

Binary variables Binary variables are those variables that can have only two unique values. For example, a variable "Smoking Habit" can contain only two values like "True" and "False."

Binning *Binning* (also called *bucketing*) is the process of converting a continuous feature into multiple binary features called *bins* or *buckets*, typically based on value range. For example, instead of representing age as a single integer-valued feature, the analyst could chop ranges into discrete bins.

Boosting Boosting is a sequential process, where each subsequent model attempts to correct the errors of the previous model. The succeeding models are dependent on the previous model.

Bootstrapping Bootstrapping is the process of dividing the dataset into multiple subsets with replacement. Each subset is of the same size of the dataset. These samples are called bootstrap samples. Bootstrapping can be used to estimate a quantity of a *population*. This is done by repeatedly taking small samples, calculating the statistic, and taking the average of the calculated statistics. *Bootstrapping* model means training a model on a small set of labeled data, and then manually reviewing unlabeled examples for errors, and then adding those to the training set in an iterative process.

Business analytics Business analytics is mainly used to show the practical methodology followed by an organization for exploring data to gain business insights. The methodology focuses on statistical analysis of the data.

Business intelligence Business intelligence ia a set of strategies, applications, data, and technologies used by an organization for data collection, analysis, and generating insights to derive strategic business opportunities.

Categorical variables Variables with a discrete set of possible values. Can be ordinal or nominal (see ordinal and nominal).

Centroid A *centroid* is the center of a cluster as determined by a *K-means* or *K-median* algorithm.

Classification It is a supervised learning method. A prediction method that assigns each data point to a predefined category, e.g., is the email spam or not spam? It can be binary classification (two classes) or multi-class classification (more than two classes).

Classification threshold It is the value that is used to classify a new observation into categories. Threshold value of probability of output is used to classify into classes.

Clustering Clustering is an unsupervised learning technique to group data into similar groups or buckets.

Confidence interval A confidence interval (CI) is used to estimate what percent of a population fits a category based on the results from a sample population.

Confusion matrix Confusion matrix is a table that is often used to describe the performance of a classification model. It is C * C matrix, where C is the number of classes. Confusion matrix is formed between predictions of model classes vs. actual classes. One axis of the confusion matrix is the *label* that the model predicted, and the other axis is the actual label. Confusion matrices can be used to calculate different performance metrics, such as *precision* and *recall*.

Continuous variables Continuous variables are those variables that can have infinite number of values within a specified range, e.g., sales, lifespan, weight.

Convergence A state reached during the training of a model satisfying certain criterion. An iterative algorithm is said to converge when as the iterations proceed the output gets closer and closer to a specific value.

Correlation Correlation is the ratio of covariance of two variables to a product of variance (of the variables). It takes a value between +1 and −1. An extreme value on both the side means they are strongly correlated with each other. A value of zero indicates a no correlation. The most widely used correlation coefficient is Pearson coefficient.

Cosine similarity Cosine similarity is the cosine of the angle between two vectors. It measures the similarity between two vectors. Two parallel vectors have a cosine similarity of 1 and two vectors at 90° have a cosine similarity of 0. Suppose we have two vectors A and B. Cosine similarity of these vectors can be calculated by dividing the dot product of A and B with the product of the magnitude of the two vectors as given below:$\text{sim}(A, B) = \cos(\theta) = \frac{A \cdot B}{\|A\| \|B\|}$

Cost function Cost function is to define and measure the error of the model. The cost function is given by:$J(\theta_0, \theta_1) = \frac{1}{2m} \sum_{i=1}^{m} \left(h_\theta(x^{(i)}) - y^{(i)} \right)^2$ where• $h(x)$ is the prediction.• y is the actual output.• m is the number of observations on the training set.

Cross validation Cross validation is a technique that involves reserving a particular sample of a dataset not used for training the model. Model is tested on this dataset to evaluate its performance. Cross validation can be used for model selection or hyper parameter tuning (see hyper parameter). There are various methods to perform cross validation such as:K-fold cross validation.Leave one out cross validation (LOOCV).Stratified K-fold cross validation.

Data frame Data frame is a two-dimensional labeled data structure with columns of potentially different types defined in the context of use of R and Python.

Data mining Data mining is a study of extracting useful information from data. Data mining is done for the purposes of market analysis, customer purchase pattern, fraud detection, predicting annual sales, etc.

Dataset A dataset (or data set) is a collection of data. A dataset is organized into some type of data structure.

Data science Data science is a combination of data analysis, algorithm development, and technology in order to solve analytical problems. The main goal is to use to generate business values.

Data transformation Data transformation is the process to convert data from one form to the other. This is usually done during preprocessing step.

Decision boundary A decision boundary or decision surface is a hyper surface that partitions the underlying vector space into two or more sets, one for each class. How well the classifier works depends upon how closely the input patterns to be classified resemble the decision boundary.

Deep learning Deep learning is associated with machine learning algorithms (artificial neural networks), which uses the concept of human brain to facilitate the modeling of some arbitrary function.

Descriptive statistics Descriptive statistics is comprised of those values, which explains the spread and central tendency of data. For example, mean is a way to represent central tendency of the data, whereas range is a way to represent spread of the data.

Degree of freedom It is the number of variables that have the choice of having more than one arbitrary value.

Dimension Dimension means how many features one has in data.

Dimension reduction Dimension reduction refers to the process of converting a set of data having vast dimensions into data with fewer dimensions ensuring that it conveys similar information concisely and consistently. Dimensionality reduction is helpful in training a *model* using a bigger dataset. Also, in many cases, the *accuracy* of the model increases after the original dataset is transformed into a dimensionality-reduced dataset.

Dummy variable Dummy variable is another name for Boolean variable derived from some other variable or a combination of variables. An example of dummy variable is that it takes value 0 or 1. 0 means value is true (i.e., gender = male) and 1 means value is false (i.e., gender = female).

Ensemble learning Ensemble learning is a problem of learning a strong classifier by combining multiple weak classifiers.

Ensemble learning algorithm An ensemble learning algorithm combines multiple weak classifiers to build a string classifier (the one with a higher accuracy than that of individual classifier).

Epoch An epoch is one that passes through the training set by a machine learning algorithm.

Exploratory data analysis (EDA) EDA is a phase used for data science pipeline in which the focus is to understand insights of the data through visualization or by statistical analysis.

Evaluation metrics The purpose of evaluation metric is to measure the quality of the ML model.

False negatives Points that are actually true but are incorrectly predicted as false.

False positives Points that are actually false but are incorrectly predicted as true.

Feature Also known as variable or attribute, is an observable quantity, recorded used to describe an object (e.g., color, size, age, weight, etc.), and is used to create a model. It can be numerical or categorical.

Feature selection Feature selection is the process of selecting relevant features from a data set that are required to explain the predictive power of an ML model. Note that it results in dropping irrelevant features.

Feature vector A *feature vector* is a vector in which each dimension represents a certain *feature* of an *example*.

F-score An evaluation metric that combines both precision and recall as a measure of effectiveness of classification.

Goodness of fit The goodness of fit of a model describes how well it fits a given set of observations capturing the discrepancy between observed value and predicted value.

Gradient descent Gradient descent is an iterative optimization technique for finding the minimum of a function. In ML algorithms, we use gradient descent to minimize the cost function. It finds out the best set of parameters.

Grid search *Grid search* is a way of *hyper parameter tuning*. The process consists of training the *model* on all possible combinations of hyper parameter values and then selecting the best combination. The best combination of hyper parameters is the one that performs the best on the *validation set* (see hyper parameter).

Holdout sample While working on the dataset, a small part of the dataset is not used for training the model instead, it is used to check the performance of the model. This part of the dataset is called the holdout sample.

Hybrid cloud When an organization maintains a local private cloud to perform on-premises computing, as well as uses servers in a public cloud for backup or overflow of computing tasks, is called a hybrid cloud.

Hyper parameter Hyper parameter is a parameter whose value is set before training an ML model. Different models require different hyperparameters and some require none. Hyper parameters should not be confused with the parameters of the model because the parameters are estimated from the data. They are tweaked to improve the performance of the model. How fast a model can learn (learning rate) or complexity of a model, and number of hidden layers in a neural networks are some examples of hyper parameters.

Hyper plane A *hyper plane* is a boundary that separates a space into two subspaces. For example, a line is a hyper plane in two dimensions and a plane is a hyper plane in three dimensions. In machine learning, a hyper plane is usually a boundary separating a high-dimensional space.

Hypothesis Hypothesis is a possible view or assertion of an analyst about the problem he or she is working upon. It may be true or may not be true.

Imputation Imputation is a technique used for handling missing values in the data.

Inferential statistics In inferential statistics, one tries to hypothesize about the population by only looking at a sample of it.

Instance A data point, row, or sample in a dataset. Another term used is observation.

Iteration Iteration refers to the number of times an algorithm's parameters are updated while training a model on a dataset.

Labeled data The output of an observation in supervised learning has a "class" or "tag" associated with each of its observation.

Learning algorithm A *learning algorithm*, or a *machine-learning algorithm*, is an algorithm that can produce a *model* by analyzing a *dataset*.

Learning rate The size of the update steps to take during optimization loops like gradient descent. During each iteration, the gradient descent algorithm multiplies the gradient by the learning rate. The resulting product is called the *gradient step*.

Log loss Log loss or logistic loss is one of the evaluation metrics used to find how good the model is. Lower the log loss better is the model. Log loss is the logarithm of the product of all probabilities. Log loss for two classes is defined as: $-(y \log(p) + (1 - y) \log(1 - p))$ where y is the class label, and p is the predicted probability.

Machine learning *Machine learning* is at the intersection of the subfields of computer science, mathematics, and statistics that focus on the design of systems that can learn from and make decisions and predictions based on data.

Maximum likelihood estimation It is a method for finding the values of parameters that maximize the likelihood. The resulting values are called maximum likelihood estimates (MLE).

ML-as-a-service (MLaaS) ML as a service is an array of services that provide machine learning tools as part of cloud computing services. This can include tools for data visualization, facial recognition, natural language processing, predictive analytics, and deep learning.

Model A mathematical representation of a real-world process, a predictive model forecast a future outcome based on past behavior. Models are created/learned during training an algorithm on a dataset.

Model selection Model selection is the task of selecting a statistical model from a set of known models.

NaN NaN stands for "not a number." It is a numeric data type representing an undefined or unrepresentable value.

Natural language processing (NLP) Natural language process, or NLP for short, is a field of study focused on the interactions between human language and computers. NLP helps machines "read" text by simulating the human ability to understand language. It sits at the intersection of computer science, artificial intelligence, and computational linguistics.

Nominal variable Nominal variables are categorical variables having two or more categories without any kind of order.

Normalization Normalization is the process of rescaling data so that they have the same scale typically in the interval $[-1,+1]$ or $[0,1]$. Normalization is used to avoid over-fitting and improving computation speed. Normalization is used when the attributes in data have varying scales.

Noise Any irrelevant information or randomness in a dataset that obscures the underlying pattern.

Observation A data point, row, or sample in a dataset.

One hot encoding One hot encoding is done usually in the preprocessing phase. It is a technique that converts categorical variables to numerical value.

Ordinal variables Ordinal variables are those variables that have discrete values but have total order involved.

Outlier An observation that deviates significantly from normal observations in the dataset.

Over-fitting Over-fitting occurs when model learns the training data too well and incorporates details and noise specific to dataset. One can tell a model is over-fitting when it performs great on training set, but poorly on test set (or new data).

Parameters Parameters are properties of training data learned by training a machine learning model. They are adjusted using optimization function and unique to each experiment.

Pattern recognition Pattern recognition is the ability to detect arrangements of characteristics or data that yield information about a given system or data set. Pattern recognition is essential to many overlapping areas of IT, including big data analytics, biometric identification, security, and artificial intelligence (AI).

Precision Precision can be measured as of the total actual positive cases; how many positives were predicted correctly. It can be represented as:Precision = TP/ (TP + FP).

Predictor variable Predictor variable is used to make prediction for dependent variable.

Private cloud If an organization maintains its compute resources in-house and shares them with internal users, it is called a private cloud. Typically, these users belong to the organization that maintains the private cloud resources.

Public cloud When cloud resources are located remotely and made available to the users, anytime and anywhere, on a commercial basis is called a public cloud.

P-value P-value is the probability of getting a result equal to or greater than the observed value, when the null hypothesis is true.

Random search *Random search* is a *hyper parameter tuning* technique in which the combinations of different hyper parameter values are first generated, and then they are sampled randomly and used to train a model.

Range Range is the difference between the highest and the lowest value of the data. It is used to measure the spread of data.

Recall Recall is described as the measured of how many of the positive predictions were correct. It can be represented as:Recall = TP/(TP + FN).

Recommendation algorithms Algorithms that help machines suggest a choice based on its commonality with historical data.

Regression A prediction method whose output is a real number, i.e., a value that represents a quantity along a line. It is a supervised learning method. Example: predicting the stock price or the revenue of a company.

Regularization Regularization is a technique utilized to combat the over fitting problem. This is achieved by adding a complexity term to the loss function that gives a bigger loss for more complex models.

Reinforcement learning Training a model to maximize a reward via iterative trial and error. *Reinforcement learning* is a subfield of machine learning where the machine perceives its environment's *state* as a vector of features. The machine can execute *actions* in every state, and different actions bring different *rewards* and also move the machine to another state. The goal of the reinforcement learning is to learn a *policy* that is the prescription of the optimal action to execute in each state. The action is optimal if it maximizes the average reward.

Residual Residual of a value is the difference between the observed value and the predicted value of the quantity of interest. Using the residual values, one can create residual plots that are useful for understanding the model.

Response variable Response variable (or dependent variable) is that variable whose variation depends on other variables (predictors or independent variables).

Receiver operating characteristic (ROC) curve A plot of the true-positive rate against the false-positive rate at all classification thresholds. This is used to evaluate the performance of a classification model at different classification thresholds.

Robotic process automation (RPA) It uses software with AI and ML capabilities to perform repetitive tasks earlier performed by humans.

Root mean squared error (RMSE) RMSE is a measure of the differences between predicted value and actually observed values. It is the standard deviation of the residuals. The formula for RMSE is given by:$RMSE = \sqrt{\dfrac{\sum_{i=1}^{N}(\text{predicted}_i - \text{Actual}_i)^2}{N}}$

where• Predicted is the predicted value.• Actual is the actual observed value.• N is the total number of data points.

Similarity metric A *similarity metric* is a function that takes two *feature vectors* as input and returns a real number that indicates how these two feature vectors are "similar." Usually, the more two vectors are similar the higher is the real number. Most often, similarity metrics are used in clustering.

Skewness Skewness is a measure of symmetry. A dataset is symmetric if it looks the same on the left and right of the center point.

Standardization Standardization (or Z-score normalization) is the process where the features are rescaled, so that they will have the properties of a standard normal distribution with mean equal to zero and standard deviation equal to 1. Z-score is calculated as follows:$z = \frac{x-\mu}{\sigma}$where μ is the mean and σ is the standard deviation.

Standard error A standard error is the standard deviation of the sampling distribution of a statistic. The standard error is a statistical term that measures the accuracy of which a sample represents a population.

Supervised learning Training a model using a labeled dataset. The goal is to predict a label from a given set of predictors. Using the set of predictors, a function is generated which maps inputs to desired outputs. The goal is to approximate the mapping so well that predicted output to the new data is close to the desired output.

Target In statistics, it is called the dependent variable, it is the output of the model or value of the variable one wishes to predict.

Tokenization Tokenization is the process of splitting a text string into units called tokens. The tokens may be words or a group of words.

Training The process of creating model from the training data. The data is fed into the algorithm, which learns a representation for the problem, and produces a model that can be used for prediction.

Training set The *training set* is a collection of randomized *examples* used as input to the *learning algorithm* to create a *model.*

Test set A dataset with the same structure as training data, used to measure the performance of a model. How generalizable is model to unseen data?

True negative These are the points that are actually false and we have predicted them false.

True positive These are the points that are actually true and we have predicted them true.

Type I error: False positives Consider a company optimizing hiring practices to reduce false positives in job offers. A type I error occurs when a candidate seems good and they hire him, but he is actually bad. The decision to reject the null hypothesis could be incorrect.

Type II error False negatives. The candidate was great but the company passed on him. The decision to retain the null hypothesis could be incorrect.

T-test *T*-test is used to compare two datasets by comparing their means.

Under-fitting Under-fitting occurs when a model over-generalizes and fails to incorporate relevant variations in the data that would give the model more predictive power. One can tell a model is under-fitting when it performs poorly on both training and test sets.

Unstructured data Information that either does not have a pre-defined data model or is not organized in a pre-defined manner.

Unsupervised learning Training a model to find patterns in an unlabeled dataset (e.g., clustering). In this type of learning, we do not have a target or outcome variable.

Validation set A set of observations used during model training to provide feedback on how well the current parameters generalize beyond the training set. If training error decreases but validation error increases, the model is likely overfitting and one should pause training.

Variance *Variance* is an error from sensitivity to small fluctuations in the *training set.•* *Low variance* suggests model is internally consistent, with predictions varying little from each other after every iteration.• *High variance* (with low bias) suggests model may be over-fitting and reading too deeply into the noise found in every training set.

Weak classifier In ensemble learning, a weak classifier is usually one of a collection of low-accuracy classifier, which, when combined by an ensemble learning algorithm, can produce a string classifier.

Z-test Z-test determines to what extent a data point is away from the mean of the dataset, in standard deviation.

Index

Printed in the United States
by Baker & Taylor Publisher Services